A COLOUR HANDBOOK OF
DISEASES OF SMALL GRAIN CEREAL CROPS

Timothy D. Murray
Professor, Washington State University, USA

David W. Parry
Head of Entomology and Plant Pathology Department,
Horticultural Research International, East Malling, England

Nigel D. Cattlin
Holt Studios International, England

MANSON
PUBLISHING

Advisory Editor

Professor G.R. Dixon
University of Strathclyde
Scotland

Copyright ©1998 Manson Publishing Ltd
ISBN 1-874545-39-1

CIP catalogue records are available from the British Library

For full details of all Manson Publishing Ltd titles please write to Manson Publishing Ltd, 73 Corringham Road, London NW11 7DL, UK.

Project management: John Ormiston

Design: edi The FDI Partnership

Colour reproduction: Tenon & Polert Colour Scanning Ltd., HK.
Printed by Grafos SA, Barcelona, Spain

Contents

1. Ear and grain diseases   1

2. Leaf and stem diseases 25

Preface

Practitioners of plant pathology need to make visual diagnoses of disease symptoms regularly; these may then be supported and confirmed by detailed cultural, physiological and molecular laboratory tests. *A Colour Handbook of Diseases of Small Grain Cereal Crops* makes the diagnosis of diseases of wheat, barley, oats, and rye easier for growers, farmers, agricultural specialists and consultants, plant scientists, those in the agrochemical and breeding industries, and students alike.

High quality photographs of diagnostic symptoms and pathogen structures are combined with text that describes the geographic distribution, pathogen symptoms, economic importance, disease cycle, and disease control. The diseases are organized first by the part of the plant affected (Ear and Grain, Leaf and Stem, and Stem Base and Root, Sections 1–3), and within these by type of disease symptom (blights, bunts/smuts, rust, mildew, and so forth). This approach enables the user readily to narrow the list of diseases under consideration.

Several graphic features are provided throughout the book to ensure it is as convenient to use as possible. Beside each disease title are cereal icons – those highlighted (in colour rather than grey) show the species affected – and maps on which worldwide distribution is shown by the shading.

To serve as an introduction to plant pathology, the basic principles are given (pp. ix–xiii), and a simple diagnostic guide is provided (pp. xv–xviii). For readers who have a ×10 hand lens or access to basic microscopic facilities, spore types of most of the major fungal pathogens are provided as an additional aid to diagnosis (Section 4). In addition, Appendix Tables (pp. 129–133) summarize, for the most important diseases, the diagnostic features, worldwide distribution, and cereals affected. A Glossary (pp. 125–128) and Bibliography (pp. 134–138) are also provided.

The authors trust that this book will not languish on library and office shelves, but will provide excellent service in the field.

Timothy D. Murray
David W. Parry
Nigel D. Cattlin

Acknowledgements

Nigel Cattlin is very grateful to the many people who, over the years, have found disease symptoms for him to photograph for this handbook. His thanks go in particular to those who have helped find some of the more elusive examples during the final stages of the book's preparation: Professor William W. Bockus (Kansas State University), Dr Jack Riesselman (Montana State University), Dr D. McVey (Cereal Rust Laboratory, University of Minnesota), Dr Roland Line (USDA, ARS Pacific West Area), Mr Andy Leadbeater (Novartis, Cambridge), Mr Arthur Marshall (Woodford Farm Services, Devon), Mr Douglas Bain (Agrochem Arbroath Ltd, Aberdeen) and, of course, his co-authors.

Timothy Murray thanks Drs G.W. Bruehl and R.F. Line for comments and suggestions during the preparation and proofreading of the manuscript. The technical assistance and proofreading skills of C.A. Blank are also gratefully acknowledged.

Inevitably, it was not possible to find and photograph every disease described, and we thank the following people who have allowed us to use their photographs: Professor Robert L. Forster, University of Idaho (illustrations 5, 6, 7, 8, 10, 134, 135, 139), CIMMYT, Mexico (illustrations 16, 17, 18), Professor Thomas W. Carroll, Montana State University (illustrations 43, 44, 45, 46), Dr Robert L. Bowden, Co-operative Extension Service, Kansas State University (illustrations 58, 59, 60, 62, 64, 65, 66), Dr R.G. Rees, Queensland Wheat Research Institute, Toowoomba (illustrations 95, 96), and Professor J. Drew Smith, University of Saskatchewan (illustrations 130, 131, 132).

Timothy D. Murray
David W. Parry
Nigel D. Cattlin

Introduction

Defining Plant Disease

Many different definitions of plant disease exist, none of which is accepted by everyone. For this book, the broad definition developed by the US National Academy of Sciences in 1968 is used. Accordingly, a plant disease is: *A harmful alteration of the normal physiological and biological development of a plant resulting in abnormal morphological and physiological changes (symptoms).* This definition includes two parts that are useful for our purposes. First, plant disease is a harmful alteration of the normal physiological processes of a plant. Clearly, we would not be interested in disease if it were not a harmful development. Second, the harmful change results in an abnormal manifestation of the plant – symptoms, visual signals that the plant is not developing normally. A significant portion of this book is devoted to the visual signals of small cereal grain diseases that aid correct diagnosis.

Three factors are required to initiate a plant disease: a suscept (the plant), a pathogen, and favourable environmental conditions. The relationship between these factors and their interdependence are often depicted as a triangle, with the pathogen, suscept, and environment at the corners of the triangle. This depiction is meant to emphasise the interdependence of these factors and their necessity to the occurrence of plant disease. However, the disease triangle does not reflect the dynamic interactions that occur between the plant, pathogen, and environment in which they come together. Disease is not an either/or process, but rather it occurs in degrees. Subtle changes in the environment, or host susceptibility and/or resistance, or pathogen virulence can change the relative severity of disease without stopping the process. For this reason, the depiction of the suscept, pathogen, and environment as circles in the 'plant disease triad' (I) reflects the dynamic interaction that occurs among these factors.

In addition to illustrating the occurrence of plant disease, the disease triad is also a convenient model in which to consider the control of plant diseases. After all, if subtle changes in the relationships between the plant, pathogen, and environment can increase the severity of disease, such changes can be used to limit development of disease too.

Causes of Plant Disease

The definition of plant disease given above does not state the causes of plant disease. According to this definition, anything that causes a harmful alteration of the normal physiological processes of a plant is a causal agent of plant disease. Biotic organisms, such as fungi, bacteria, phytoplasmas, viruses, viroids, nematodes, and parasitic plants, are usually considered as causal agents of plant disease. However, abiotic causes of plant disease also occur. Indeed, mineral deficiencies cause harmful changes in the physiology of plants that result in observable symptoms. One significant distinction between these causal agents of plant disease is the fact that abiotic entities do not spread from one affected plant to another; in other words, they

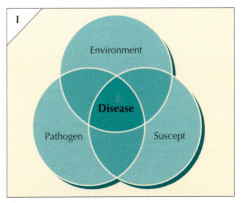

▲ The plant disease triad. Subtle changes in the environment, pathogen, or suscept can influence the severity of the disease.

are not transmissible. In contrast, biotic causes of plant disease are transmissible – they spread from one infected plant to another. The latter point is significant when considering disease control options. This book focuses on the transmissible pathogens of the small cereal grains caused by fungi, bacteria, and viruses.

All biotic pathogens of plants derive nutrition from the plants they infect. However, not all pathogens have the same nutritional relationship with their respective plant hosts. Variation in nutritional strategies is especially evident among the fungi, but also occurs to some extent among the bacteria. Pathogens that utilize dead plant material as a food source are called *saprophytes*, whereas those that use living plants are *parasites*. Pathogens that live only on a living plant host are called *obligate parasites*, whereas those that can live on both living and dead plants are called *facultative parasites* or *facultative saprophytes*, depending upon the relative importance of the living and dead plants to the organism. Obligate parasites are sometimes referred to as *biotrophs* and facultative parasites and saprophytes as *necrotrophs*. These distinctions between nutritional strategies are important when considering disease control strategies, especially for soil-borne fungi. For example, facultative saprophytes (more reliant on living host plants) depend upon host residue for survival and may be effectively controlled by crop rotation, whereas facultative parasites (more reliant on dead plant material) can survive as saprophytes in soil and are not effectively controlled by crop rotation.

The Disease Cycle

The disease cycle is a concept used by plant pathologists to describe the sequence of events involved in development of disease caused by biotic pathogens, including the appearance, development, and perpetuation of disease, as well as the survival of the pathogen between susceptible hosts. In many ways, a disease cycle is like a life cycle, which describes the development of a single organism. For example, the life cycle of an annual plant begins when the seed germinates and is followed by vegetative growth of

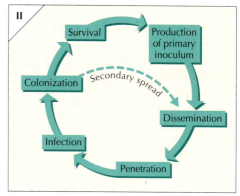

▲ An idealized disease cycle showing the stages involved in disease development.

the plant, reproductive growth (during which flowers and seed are produced), and finally death of the plant. The seed is disseminated and germinates when conditions become favourable, thus completing the life cycle. In contrast, a disease cycle describes the interaction of two organisms, the pathogen and the plant, which results in the development of disease. Just as the life cycle of an individual organism is divided into distinct phases, disease cycles are divided into distinct stages that describe significant events in the development of a disease (**II**).

Viewing plant diseases from the perspective of the disease cycle allows for a greater understanding of the factors important to the development of individual diseases and, as such, may offer insights to its control. In addition, understanding a disease cycle enables predictions to be made about its occurrence within regions and years.

The discussion that follows describes stages in an idealized disease cycle and includes those stages that are common to all diseases caused by transmissible pathogens. That is not to say, however, that each stage is easily recognized or understood for all plant diseases. Indeed, there is a great deal that is not understand about many diseases.

Production of inoculum

Inoculum is any part of the pathogen that is capable of infecting a plant. The types of inoculum produced depend upon the particular pathogen. With fungi, for example, inoculum is often various types of spores, but may also be fragments of hyphae contained within colonized host debris. With bacteria and viruses, the entire bacterial cell or virus particle are the inocula. Two types of inoculum are recognized, depending upon when and where it is produced. *Primary inoculum* is produced from survival structures and initiates the disease cycle within a particular growing season. It results in *primary infections. Secondary inoculum* is produced during the current growing season on infected plants and is a result of primary infections. It results in secondary infections and secondary disease cycles. Not all pathogens produce secondary inoculum.

Dissemination

Dissemination is the spread of inoculum from the location where it is produced to the plant. The most common vehicles for the spread of plant pathogens are wind and water. Other methods of spread include *vectors*, such as insects, and human activities. Inoculation is the term used to describe the initial contact of the pathogen and the plant. Inoculation does not imply disease development.

Penetration

Penetration is the initial entry of a pathogen into a plant. The outer covering of a plant is its first line of defence and, thus, is a significant barrier to plant pathogens. Pathogens penetrate plants in three different ways. *Direct penetration* occurs when the pathogen enters the suscept through the epidermis; it is the result of both mechanical pressure and enzymatic degradation of the outer cell wall layers. *Indirect penetration* occurs when the pathogen enters the plant through a natural opening, such as a stomate or hydathode. Many pathogens penetrate plants through *wounds* or breaks in the outer covering.

Although essential to disease development, penetration of a plant by a pathogen does not guaranty that disease will develop. Up to this point in the disease cycle, the energy used by the pathogen has been derived from its own stored sources, in the case of fungi and bacteria, and little or no damage to the plant is apparent.

Infection

Infection occurs when the pathogen contacts the internal tissues and establishes a pathogenic relationship with the plant. Another way to think of this is that the pathogen begins to obtain nutrients or derive energy for continued growth and development from the plant. The *infection court* is the specific location on the plant where infection occurs; it varies depending upon the pathogen and type of disease. Infection is the stage at which symptoms begin to develop. The *incubation period* is the time between infection and appearance of the first symptom.

Colonization

This stage is also referred to as growth and reproduction, and involves active growth and reproduction of the pathogen within the plant. The pathogen increases in size and/or number, and the amount of plant tissue occupied by the pathogen increases. Symptoms become most obvious during the colonization stage. It is also during the colonization stage that secondary inoculum is produced, which may lead to the development of secondary disease cycles. Diseases with multiple cycles per growing season are referred to as *polycyclic diseases*. Not all diseases have secondary cycles: diseases that have a single cycle per growing season are referred to as *mono-* or *single-cycle diseases*.

Survival

Pathogens must survive during periods of unfavourable environmental conditions when susceptible plants are unavailable if a disease is to occur in the subsequent growing season. Many pathogens, especially fungi, have specialized structures for survival that are produced during the colonization stage. However, many other pathogens, such as bacteria, viruses, and some fungi, do not have specialized survival structures. These pathogens may survive in association with host debris, in association with perennial plants, or with insect vectors. Depending upon when survival occurs, it may be described as *over-wintering* (survival in the winter) or *over-summering* (survival in the summer).

Control of Plant Disease

Control, which is our ultimate goal, is the application of practices devised to reduce the damage or loss attributable to plant disease. Control measures must be cost-effective, particularly when the direct costs of application of fungicides are being considered. However, reliable relationships between disease severity and yield losses are scarce and hence the economic importance of individual diseases given in the text is often very tenuous, unless based on substantial scientific data.

There are two prerequisites for effective disease control, of which the first is accurate diagnosis of the disease problem. Control measures directed against one disease may not be effective against another, and without an accurate diagnosis there is no way to be sure of the disease problem. Second is the timely application of disease control practices. *When* a disease control practice is implemented can be as important as the practice itself. In general, disease control practices applied early in the disease cycle provide better control than those applied later.

When the application of a fungicide is being considered, it may be prudent to use disease threshold criteria to assist with the decision. Such criteria attempt to use disease cost–fungicide benefit relationships to establish a permitted disease severity. If the disease threshold is exceeded, then application of a fungicide should be economically viable. Such thresholds usually occur when 1–5% of the leaf area is affected, depending upon the disease and the fungicide product recommendations. Unfortunately, as previously mentioned, relationships between disease severity and yield loss are often unreliable. In addition, other factors such as weather and varietal susceptibility may push the disease very quickly beyond the threshold. Hence, it is wise not to rely too heavily on such disease thresholds or to use them in isolation.

All methods of disease control attempt to modify the plant disease triad to make conditions unfavourable for disease development. Different control practices target different parts of the disease triad; knowing which practice to use requires an understanding of the disease cycle. All disease control practices can be divided into one of the following four categories.

Exclusion

Exclusion is to prevent a pathogen from entering or becoming established in an area in which it does not occur – the area may be an individual field or a country. Most exclusionary disease control practices involve legal means of regulating the movement of agricultural commodities (including plants, plant parts, and soil), such as quarantines, embargoes, inspections at ports of entry, and phytosanitary certification of plant materials before shipment.

Eradication

Eradication is to eliminate a pathogen from an area in which it has already become established. Complete elimination is expensive and seldom accomplished, except in situations that involve recent introductions where the pathogen has not become fully established. Specific control practices for eradication are often further subdivided into cultural and sanitation, physical, and chemotherapy categories. Cultural and sanitation methods of disease control are those practices associated with crop husbandry and post-harvest handling of the crop, respectively, and include activities such as crop rotation, tillage, removal of infected plants or plant parts, irrigation and fertility management, and the use of pathogen-free seed. The use of heat, radiation, or chemicals, such as soil fumigants, are all considered physical methods of disease control. Treatments of infected plants with systemic pesticides that have eradicant properties or with antibiotics are considered chemotherapy.

Protection

Protection is to establish a protective barrier between the plant and pathogen to prevent infection. Historically, protection has been mediated through the application of pesticides, but more recently biological control agents have also been used in a few cases to protect plants from pathogens.

Disease Resistance

Disease resistance is the development of plant populations that are immune, highly resistant, or tolerant of the activities of the pathogen. Disease resistance is perhaps the most desirable form of disease control, since it is easy to implement, environmentally acceptable, and affordable. However, resistance is not available for all pathogens and for many pathogens resistance is not stable.

Summary

Often the most successful and cost-effective disease control programmes use more than one approach; that is, they use integrated disease control. It is particularly important to consider an integrated approach to disease control where the risk of fungicide resistance in a pathogen population is high.

Specific control recommendations for a particular disease may change and can vary among countries and also among production areas within countries, especially when cultivars and pesticides are involved. For this reason, readers are encouraged to consult with the local extension or crop advisory personnel for the most current recommendations in the area or country of interest.

Diagnostic guide for plant diseases

This book is designed to assist readers in the visual diagnosis of cereal diseases. In order to carry out this task effectively, as many as possible of the following questions should be answered.

1. What is the crop under investigation – wheat, barley, oats, or rye? Some diseases will only affect specific crops.
2. What is the variety of the crop?
3. Is this variety of the host known to be especially sensitive to particular diseases?
4. Which part or parts of the crop are most affected with the disease – the roots, stem, base, leaves, or heads and/or ears? This book is organized by afflicted plant parts, so it is possible to identify many diseases by looking through the chapter on affected parts.
5. What are the key diagnostic symptoms? It is essential to compare a healthy plant with the afflicted part to identify correctly the key symptoms. Also, be aware that symptoms of plant disease change over time, so it is best to examine many plants over time to identify the diagnostic symptoms. Consider the following:

- Is there rapid and comparatively extensive death of tissue, usually characterized by bleaching and/or blackening of the foliage and heads and/or ears? If yes, the disease is a blight (**A**).
- Are there patches of superficial light-coloured mould growing on the plant surface which when disturbed release masses of dust-like spores? If yes, the disease is a powdery mildew (**B**).
- Are the heads and/or ears or leaves covered with a dark, furry or granular mould, espe-

A. Diagnostic symptoms for blight.

B. Diagnostic symptoms for powdery mildew.

cially after periods of high humidity? If yes, the disease is a sooty mould (**C**).

- Are there small, rust coloured (yellow or brown) pustules containing tiny pollen-like spores on the plant surface? If yes, the disease is a rust (**D**).

- Are the grains completely replaced by a mass of dark, powdery spores that are either encased (bunt, **E**) or exposed to the air (smut, **F**), or are there long, dark-coloured pustules on the leaves? If yes, the disease is a smut (**F**).

C. Diagnostic symptoms for sooty mould.
D. Diagnostic symptoms for rust.
E. Diagnostic symptoms for bunt.

- Are there spots or blotches on leaves that are initially yellow-coloured and then turn brown? The spots or blotches initially may be small, but enlarge, coalesce, and cover large areas. If yes, the disease is a leaf spot or blotch (**G**).

- Is there a general slimy rot of the plant tissues following prolonged snow cover with either dark, speckled structures (snow scald or speckled snow mould) or a pink or orange growth on the tissue (pink snow mould)? If yes, the disease is a snow mould (**H**).

F. Diagnostic symptoms for smut.
G. Diagnostic symptoms for leaf spot or blotch.
H. Diagnostic symptoms for snow mould.

- Is there a thin yellow stripe running parallel to the veins along the leaf? If yes, the disease is a stripe (**I**).
- Are the leaves mottled with irregularly shaped yellow, green, and/or white patches? If yes the disease may be a mosaic (**J**).

6. Are there physical structures produced by the pathogen, such as small black spore cases? If so, these can be extremely useful in positively identifying the causes of the disease.

7. Were there any unusual environmental conditions (excessive heat or cold, hail, floods, drought, etc) that preceded the appearance of the symptoms? Did they appear suddenly, uniformly, and over a relatively large area of the field? If yes, the disease is likely to be abiotic in origin and will not spread.

I.   Diagnostic symptoms for stripe.

J.   Diagnostic symptoms for a mosaic.

Ear and Grain Disease

Blights

Fusarium avenaceum (Gibberella avenacea), F. culmorum, F. graminearum (Gibberella zeae), F. poae, Microdochium nivale (formerly F. nivale) (Monographella nivalis)

Diseases: Fusarium Seedling Blight, Foot Rot, and Head (Ear) Blight

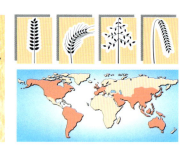

Each of the three Fusarium diseases of cereals may be caused by any one of the above pathogens. In temperate regions, *F. culmorum* and *M. nivale* predominate, whereas in hotter climatic areas, *F. graminearum* is more common. All cereals can be affected by the diseases, but research has focused on wheat. Severe symptoms of Fusarium seedling blight consist of pre- and post-emergence damping off. Should seedlings survive, they often develop brown lesions around soil level. It is also possible for symptomless infections to occur.

Fusarium foot rot (crown rot, brown foot rot) symptoms are varied. In temperate areas, the most common symptom is a dark brown lesion around the node of mature plants. Long, thin, dark brown vertical streaks are also frequently observed. In more arid areas, dryland foot rot may develop. The entire stem base becomes girdled with a dark brown lesion. Tissue may become soft and white, or pink fungal growth with orange spore masses can develop.

Initial symptoms of Fusarium ear blight consist of small water-soaked brownish spots at the base or middle of the glume. Water soaking and discoloration then spread in all directions from the point of infection (**1**). Premature death or bleaching of cereal spikelets is also a common symptom, and in humid conditions white or pink fungal growth with orange spore masses may develop (**2, 3**). Symptoms of Fusarium ear blight caused by *F. graminearum* can also include a scabbed

appearance as the result of blue-black perithecial development under humid conditions.

Disease cycle

The epidemiological relationship between the three Fusarium diseases has yet to be elucidated fully.

1. Ear blight with bleached grains on ears of infected wheat.

2. Bleached wheat grains with orange ear blight sporulation.

3. Wheat ear with orange sporulation and white mycelium of ear blight.

All of the species implicated in the disease can survive saprophytically in the soil or on plant material of a range of different crops and weed species. In addition, they can all be seed-borne on cereals. Seedling blight caused by *M. nivale* tends to be most severe under cool, wet soil conditions, whereas warmer, drier soils are more conducive to seedling blight caused by *F. avenaceum*, *F. culmorum*, and *F. graminearum*. It is possible that seedlings which survive the initial infection may develop foot rot at a later stage of growth. Again, environmental conditions are likely to influence disease development, with moisture stress resulting in severe symptoms of dryland foot rot caused by *F. culmorum* and *F. graminearum*. Under such conditions, sporulation may occur on stem bases and nodes. Spores may be rain-splashed directly to ears or reach ears by a series of leaps involving leaves. Ascospores of *Monographella nivalis* are released after periods of high humidity and are wind-blown to ears. Wheat ears are most vulnerable to infection during anthesis and there is evidence that pollen stimulates germination of spores. If prolonged humid weather persists after infection, severe Fusarium ear blight may occur. Secondary infection is possible from early disease outbreaks, although wheat ears become much more resistant after anthesis. Fusarium ear blight can result in contaminated grain which, if harvested and used as seed, completes the disease cycle.

Economic importance

Fusarium seedling blight can have a significant effect if heavily contaminated seed is sown without treating with fungicide: emergence may be reduced by 80%.

The economic importance of Fusarium foot rot is difficult to determine for three reasons. First, the effect of each individual pathogen on yield may vary. Second, naturally occurring foot rot may be the result of multiple infection by two or more *Fusarium* species alongside other important stem base pathogens, such as *Pseudocercosporella herpotrichoides* (eyespot). Finally, as yet there are no highly effective and reliable fungicides to treat the disease. Recent unpublished work has shown that foot rot caused by *F. culmorum* and *M. nivale* may reduce yield by over 30%.

It is also difficult to assess accurately the effect of Fusarium ear blight on yield for reasons similar to those given for Fusarium foot rot. The regression of yield reduction (y) on Fusarium head blight (x) has been shown to vary over years and ranges from $y = 6\sqrt{x}$ (in 1987) to $y = 7.2\sqrt{x}$ (in 1989). Several of the *Fusarium* species, including *F. culmorum, F. graminearum,* and *F. poae* can, under certain conditions, produce harmful mycotoxins; this may be more important than yield reduction in many cases. The most significant mycotoxins produced by the cereal *Fusarium* species the tricothecenes, including deoxynivalenol and zearalenone. Such compounds can affect feed intake in pigs and fertility in a range of farm animals. Their effect on human health is not well-documented, but several countries (including Canada and the United States) have set maximum permitted concentrations of deoxynivalenol in grain for human consumption.

Control

Cultural control of the Fusarium diseases includes disposal of contaminated debris and crop rotation. Fusarium ear blight has been shown to be severe following maize.

Genetic resistance to Fusarium ear blight, in particular, has been sought for many years in wheat-breeding programmes throughout the world. However, there are currently no varieties that are immune to the disease, although some possess a level of tolerance.

Chemical control of the Fusarium diseases has focused primarily on seedling blight, with some success. Although most isolates of *M. nivale* in the United Kingdom are resistant to the benzimidazole (MBC) fungicides, this group may still suppress disease caused by *Fusarium* species and is widely used in seed treatments, alongside azole compounds. A new group of fungicides, the phenolpyrroles, are also proving effective in controlling Fusarium seedling blight. Chemical control of both Fusarium foot rot and Fusarium ear blight is inconsistent. Again, benzimidazoles and azoles are sometimes used.

Xanthomonas campestris pv. *translucens* (syn. *X. translucens*)

Diseases: Black Chaff (Bacterial Stripe, Bacterial Leaf Streak)

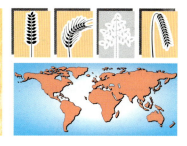

4. Discolouration on a mature wheat ear due to black chaff.

Black chaff is the most widely distributed bacterial disease of the small grains, occurring in every major cereal growing region of the world and on all of the small cereal grains, except oats. Black chaff is most severe in subtropical and tropical climates that have abundant rainfall during the growing season or where overhead irrigation is used. The common name for this disease comes from the darkened glumes of infected plants (**4**), which can be confused with glume blotch (*Septoria nodorum*) and genetic melanism. The presence of a cream-to-yellow ooze on the surface of infected plant parts (**5, 6**) during humid weather is characteristic of black chaff and distinguishes it from other diseases. Ooze is initially viscous, but dries and becomes brittle. It may appear as discrete droplets or as a thin sheet.

5. Bacterial exudate (ooze) and lesion on a barley leaf with black chaff (courtesy of Professor Robert L. Forster, University of Idaho).

Leaf symptoms begin as small, water-soaked spots or streaks (**7**) that enlarge and become translucent. Lesions are irregularly elongate and may extend the length of the leaf blade (**8**), but do not normally extend down the sheath. Lesions become necrotic with age (**9**), covered with ooze, brittle, and may resemble barley stripe mosaic or barley stripe diseases. A dark-brown to purple discoloration may also occur on the peduncle below infected ears (**10**). Awns may exhibit a 'barber's pole' symptom in which dark lesions are separated by apparently uninfected green tissue.

Disease cycle

The most important source of primary inoculum for black chaff is infected seed. The pathogen is most common on the outer surface of the seed, but may also reside within the seed. Epiphytic bacteria on volunteer crop plants or grassy weeds can also serve as inoculum sources. Bacteria are spread by splashed water, plant-to-plant contact, and aphids. Transmission is most effective when free water is present. Penetration of the host occurs through stomata or wounds and is followed by reproduction in the intercellular spaces. Droplets of bacterial cells exude (ooze) onto the plant surface during periods of high humidity and serve as secondary inoculum. The pathogen spreads and infects glumes and kernels after ear emergence.

Black chaff occurs under a wide range of temperature and moisture conditions, but is most important in warm and wet climates or when overhead irrigation is used in crop production.

6. Bacterial exudate (ooze) on wheat stems with black chaff (courtesy of Professor Robert L. Forster, University of Idaho).

7. Water-soaked spots on a wheat seedling with black chaff (courtesy of Professor Robert L. Forster, University of Idaho).

Economic importance

Loss in grain yield is due to a reduction in kernel size and ranges up to 40%. In addition, severely infected seed may reduce germination.

Control

The most effective control for black chaff is the use of certified, pathogen-free seed. Control of volunteer crop plants and grassy weeds in and surrounding production fields

8. Necrosis of flag leaves of wheat with black chaff (courtesy of Professor Robert L. Forster, University of Idaho).

9. Bacterial exudate (ooze) and lesions on leaf of barley with black chaff.

10. Dark lesions on peduncles of wheat with black chaff. (courtesy of Professor Robert L. Forster, University of Idaho)

also reduces primary inoculum. Avoiding sprinkler irrigation or managing irrigation water such that the plant canopy dries completely between irrigations reduces spread of the pathogen.

Highly effective levels of resistance are not available with current cultivars; however, cultivars that are highly susceptible should be avoided.

Treatment of seed and/or leaves with bactericides is ineffective for disease control on a large scale. Heat treatment of seed or treatment with acidified cupric acetate can produce small quantities of pathogen-free seed.

Bunts/smuts

***Tilletia controversa* – Dwarf bunt; *T. tritici* (syn. *T. caries*), *T. laevis* (syn. *T. foetida*) – Stinking smut**

Diseases: Stinking Smut (Bunt) and Dwarf Bunt

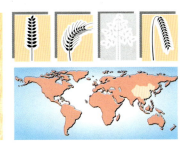

Stinking smut is an historically important disease that occurs worldwide, primarily on winter wheat. The geographic distribution of dwarf bunt is limited to areas where winter cereals are grown with prolonged snow cover including the United States, Canada, Europe, central Asia, Argentina, and Uruguay. Stinking smut derives its name from the strong odour associated with infected kernels, a symptom that is also present with dwarf bunt. The odour is due to the presence of trimethylamine, an organic compound, which in appropriate concentrations can result in explosions in combines and elevators where grain is stored. The name bunt comes from the infected seeds (bunt balls), which look like kernels, but are filled with the black, dusty teliospores of the fungus.

The diseases caused by *T. tritici* and *T. laevis* are identical, but the fungi differ in spore morphology: *T. tritici* has a reticulate pattern on the teliospores, whereas teliospores of *T. laevis* are smooth. *Tilletia controversa* is closely related to *T. tritici* and was once considered a strain of the latter fungus. The surface reticulations on spores of *T. controversa* strongly resemble those of *T. tritici*, making it difficult to distinguish these fungi by visual examination. In addition to wheat, rye, and triticale, several wild and cultivated grasses are also infected by *T. tritici* and *T. laevis*. Barley is also a host for *T. controversa*.

Obvious symptoms of these diseases are not apparent until after stem elongation begins. Plants with stinking smut may be stunted slightly, whereas those with dwarf

bunt attain only one-quarter to one-half the height of healthy plants (**11**), hence the name dwarf bunt. Ears of plants with either disease remain green longer than healthy plants, and the glumes and awns spread apart exposing

11. Wheat plant stunted by dwarf bunt compared with healthy plants in the background.

the bunt balls (**12–15**). The latter symptom may be more pronounced with dwarf bunt. Bunt balls resemble kernels, but are more rounded and have a dull, grey–green colour.

Teliospores are often released when bunt balls are ruptured during harvest; however, some remain intact and are found amongst the harvested grain.

12. Wheat plants infected with dwarf bunt in which spore masses have replaced the grain.

13. Wheat ears infected with stinking smut (no external symptoms).

14. Wheat ear before maturity; the grains are cut open to show spore masses of stinking smut.

15. Mature wheat ear where dark spore masses of stinking smut are revealed.

Disease cycle

The disease cycles of stinking smut and dwarf bunt are similar, differing primarily in when infection occurs. These pathogens survive as teliospores in soil or on seed. Seed-borne teliospores are the primary source of inoculum for stinking smut in most parts of the world. Soil-borne teliospores are important in areas with arid summers, where dry soil allows the teliospores to survive until emergence of the winter cereal in the autumn. Teliospores on or near the soil surface are the most important source of inoculum for dwarf bunt; seed-borne spores are insignificant. Germination of teliospores involves the formation of 8–16 primary basidiospores (sporidia), which fuse and in turn produce secondary basidiospores. Secondary basidiospores germinate on or near the host plant and penetrate it directly. The mycelium within the plant grows toward the apical meristem and remains there until the ovaries develop. Hyphae of the pathogen replace the tissues of the young ovary, which is then converted into teliospores. Bunt balls broken during the harvest operation release teliospores that contaminate healthy grain and are wind-disseminated to adjacent fields and contaminated soil.

Teliospores of *T. tritici* and *T. laevis* germinate when soil temperature is 5–15°C and the soil is relatively moist. The principal site of penetration and infection for stinking smut is the coleoptile. Teliospores of *T. controversa* germinate over a 3–10 week period at 3–8°C. Most germination occurs from December to February in northern latitudes and may be stimulated by snow cover. Penetration and infection by *T. controversa* occurs through the young stem near the soil surface. Teliospores of *T. controversa* in intact bunt balls can survive up to 10 years in soil. In contrast, teliospores of *T. tritici* and *T. laevis* do not usually survive for longer than 1 year in soil.

Economic importance

The smut diseases cause losses in both yield and quality. Loss in grain yield is approximately equivalent to the percentage of ear-bearing stems with smut. Even a low incidence of smut can result in grain being graded 'smutty', which brings a lower price due to the unfavourable smell and flavour associated with teliospores in the final product. In addition, some countries have quarantines prohibiting import of grain containing teliospores of *T. controversa*. The strong resemblance between teliospores of *T. tritici* and *T. controversa* has exacerbated this problem.

Control

Chemical control of stinking smut is achieved by treating seed with carboxin, some benzimidazoles, and, more recently, difenoconazole. A seed dressing of difenoconazole provides nearly complete control of dwarf bunt, but carboxin is ineffective.

Resistant cultivars may be used to control both smut diseases. However, all three pathogens are highly variable and races of each are present. Race-specific resistance is effective, although the combination of resistance and fungicide seed treatments is necessary to control stinking bunt in areas where soil-borne inoculum is important. Elsewhere, resistant varieties or seed treatments alone are adequate for control. General (race non-specific) resistance to these pathogens is not known.

Cultural practices such as shallow sowing, sowing of pathogen-free seed, and sowing when soil temperatures are unfavourable for germination of teliospores of *T. tritici* and *T. laevis* may provide partial control of stinking smut. Dwarf bunt is favoured by compacted soil and shallow sowing; thus, limiting cultural operations that compact soil and sowing seed deeply (6–8 cm) will partially control disease. Plants at the 2–3 leaf stage are most susceptible to dwarf bunt; therefore, very early or very late sowing may limit disease, but will not provide complete control.

Tilletia indica (syn. Neovossia indica)
Disease: Karnal Bunt

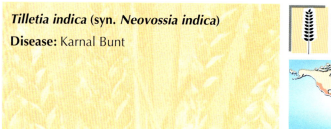

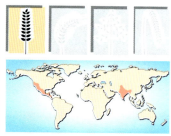

Karnal bunt, also known as partial bunt, was first described in 1931 from Karnal, Punjab, India. The disease was considered minor until an epidemic in 1970. Since that time Karnal bunt has been recognized throughout northwest India, the adjacent areas of Afghanistan and Pakistan, and parts of Nepal, Iraq, and Mexico. The disease was discovered in the southwestern part of the United States in 1996, but the extent of its distribution in the country is not yet known. The pathogen infects common and durum wheat, triticale, and several wild relatives of wheat; however, common wheat is the most susceptible host.

Symptoms of Karnal bunt are not apparent until dough stage or later and not all kernels on an individual ear are infected. Infected kernels have a darkened lesion (sorus) near the embryo end of the kernel, which is grey initially, but becomes black as the teliospores within mature (**16**). The pericarp often ruptures during harvest exposing the teliospores as a black, dusty mass. Usually only part of the kernel is affected by the disease (hence, partial bunt); however, the entire kernel may become smutted in severe cases (**17, 18**). In such cases the lesion spreads upward along the crease of the kernel into the endosperm. The fishy odour of trimethylamine found in stinking smut and dwarf bunt is associated with kernels affected by Karnal bunt. Partially smutted kernels may retain viability and produce healthy plants after germination. Minor infections of Karnal bunt may be confused with black point.

Disease cycle
Teliospores on or near the soil surface germinate by forming a germ tube containing up to 200 primary basidiospores. Primary basidiospores may be disseminated or produce secondary basidiospores that are disseminated by wind or splashing water to nearby ears of wheat. The basidiospores germinate and infection hyphae penetrate the stigma, ovary wall, or glume directly and grow into the young kernel. Disease remains localized in the kernel and does not develop systemically. Individual cells of the fungal hyphae are converted to teliospores as the kernel matures and are released when the sorus ruptures at harvest. Teliospores are disseminated by wind or as surface contaminants on harvested grain and can remain viable for up to 4 years in soil.

Environmental conditions during flowering and susceptibility of the host determine the severity of Karnal bunt. Temperatures in the range of 18–22°C (minimum of 8–10°C) and free moisture are optimal for germination of teliospores and growth of the fungus. Relative humidity greater than 70% and cloud cover with frequent showers are most favourable for disease development. Cultural practices such as overhead irrigation and over-fertilization contribute to increased disease.

Economic importance
Losses in grain yield are relatively minor even in years when epidemics occur. Reduced flour quality represents the most significant loss. Even low percentages of smutted kernels in a seed lot lead to darkening of the flour and the presence of a disagreeable odour. In addition, infected kernels may be shrunken, and have reduced test weight (specific weight) and germination.

Control
Karnal bunt is extremely difficult to control. Regulatory control measures such as embar-

16. Wheat grains with 10% karnal bunt showing disease development along the crease (courtesy of CIMMYT, Int).

17. Wheat grains severely affected by karnal bunt (courtesy of CIMMYT, Int).

18. Wheat grains severely affected by karnal bunt (courtesy of CIMMYT, Int).

against the importation of wheat seed in countries where Karnal bunt occurs is practised by many countries.

Cultural practices, including rotation away from wheat for 2 years, irrigation management, and avoiding over-fertilization, may help reduce disease. Burning straw or solarizing soil by mulching with a polyethylene tarp to raise soil temperature reduces viability of teliospores buried up to 10 cm deep in the soil.

Disease resistance offers great promise for control of Karnal bunt. Several highly resistant wheat lines from Brazil, China, India, Italy, Mexico, and the United States have been identified and are being used in breeding programmes.

Several azole and benzimidazole fungicides applied at ear emergence can reduce disease incidence. These and several other fungicides have been shown to reduce the germination of teliospores. However, it is unlikely that these fungicides will persist long enough in plants to reduce the infection.

Ustilago avenae (syn. *U. nigra*) – **Semi-loose smut;** *U. segetum* (syn. *U. kolleri*), *U. hordei* – **Covered smut**

Diseases: Semi-Loose Smut and Covered Smut

Semi-loose smut, also known as black loose smut and false loose smut, is a disease of barley and oats that is distributed worldwide. Covered smut, which affects oats and barley primarily, but also occurs on rye and wild grasses, is probably more widespread and significant than true loose smut and semi-loose smut combined. Accurate estimates of the distribution and significance of semi-loose smut are lacking, however, owing to the inability to distinguish it from true loose smut in the field based on symptoms.

Symptoms of these diseases are not apparent until the inflorescence emerges from the boot. Typically, the inflorescence of infected stems emerges at the same time as, or slightly later than, that of healthy stems. All of the florets of an infected inflorescence are replaced by masses (sori) of dark brown to black spores (**19–22**). Semi-loose and covered smut differ in the persistence of the membrane covering the sori. Persistence is determined by cultivar and strain of the pathogen: the peridium in semi-loose smut is more delicate and ruptures sooner (**19, 20**) than that of covered smut, which persists until harvest (**21, 22**). In addition, sori may develop on leaf blades and stem nodes with covered smut. All floral parts except the rachilla or rachis may be colonized by these pathogens, although the awns, lemma, and palea usually remain intact with covered smut.

19. Smutted and empty oat ears affected by semi-loose smut.

20. Oat ears affected by semi-loose smut with exposed spore masses.

21. Barley ears with covered smut.

22. Barley ears with covered smut.

Disease cycle

Semi-loose smut and covered smut have disease cycles similar to stinking smut. The primary source of inoculum is teliospores that survive as contaminants on the surface of kernels or under the lemma and palea, although soil-borne teliospores can be important in arid areas where winter barley is grown. Teliospores germinate at the same time as the host seed by producing a germ tube (promycelium) and primary basidiospores. The primary basidiospores may fuse or produce secondary basidiospores that fuse and produce the infection hyphae, which penetrate the coleoptile directly. An alternative mode of infection occurs with semi-loose smut, wherein teliospores under the lemma and palea germinate and penetrate the superficial tissues of the developing kernel, but then remain dormant until the seed germinates. The hyphae resume growth when the seed germinates, and penetrate the coleoptile. Once through the coleoptile, the fungus colonizes tissues near the apical meristem, resulting in a systemic infection. Floral organs are colonized and replaced by fungal hyphae when they form. Cells of the hyphae develop into teliospores as the inflorescence emerges. Teliospores of semi-loose smut are liberated from anthesis through harvest; however, teliospores are liberated at harvest with covered smut. Dissemination of both pathogens occurs by way of direct contact and wind.

Relatively dry soils with temperatures of 15–21°C are optimal for development of semi-loose smut. In contrast, moist to wet soils with temperatures of 20–24°C are optimal for development of covered smut.

Economic importance

Losses in grain yield are approximately equivalent to the percentage of stems with infected ears. In addition, grain graded 'smutty' receives a dockage. As with other smut diseases, even relatively low percentages of smut in a field can lead to smutty grain. Covered smut is more important in the United States and Canada than semi-loose smut.

Control

Treatment of seed with the systemic fungicide carboxin is the primary method of control for semi-loose and covered smuts, and is very effective.

Cultural practices including the use of certified pathogen-free seed effectively control these diseases. Crop rotation will control semi-loose smut in areas where soil-borne inoculum is significant because teliospores do not survive long periods of time in soil.

Disease resistance is available and has been used effectively for control. Races of the pathogen exist, but are apparently of little consequence in most areas.

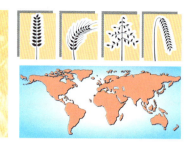

***Ustilago tritici* – wheat; *U. nuda* – barley; *U. avenae* – oats**

Disease: Loose Smut

Loose smut occurs wherever wheat, barley and oats are grown. The disease is most prevalent in areas with high humidity and rainfall in the spring during anthesis. Wheat and barley are equally susceptible to these fungi.

Symptoms of loose smut are not apparent until ear emergence. Ears of infected plants emerge earlier, have a darker colour, and are slightly taller than ears from healthy plants. All spikelets of infected ears are transformed into masses (sori) of dry, olive-black teliospores (**23–26**). Initially, the sori may be covered by a delicate, light-coloured membrane, but this soon ruptures, releasing the teliospores. Within a few days of emergence the teliospores are gone leaving, an empty rachis. Infected kernels have no visible symptoms and are fully germinable.

Disease cycle
This fungus survives as dormant hyphae in the embryo of infected seed. Following germination of the seed, the mycelium grows toward and colonizes tissues near the apical meristem, the culm nodes, and floral primordia, resulting in a systemic infection. Hyphae of the pathogen colonize and replace the ovaries, lemma, and palea as they form. Cells of the hyphae are converted into teliospores as the plant nears ear emergence. Teliospores are

23. Wheat ears with exposed spore masses of loose smut.

24. Wheat ears with exposed spore masses of loose smut.

25. Barley ears with exposed spore masses of loose smut.

26. Barley ears with exposed and partially exposed spore masses of loose smut.

wind-disseminated to open flowers of nearby plants where they germinate. Hyphae penetrate the ovary or stigma directly and grow toward and colonize the embryo and scutellum of the developing seed.

Environmental conditions during anthesis have a strong influence on disease incidence in the subsequent crop. Loose smut is favoured by frequent rain showers, high humidity, and temperatures in the range 16–22°C. However, heavy rains during anthesis can reduce disease incidence. Free-water is necessary for germination of teliospores and penetration, and cool temperatures prolong the flowering period.

Economic importance

Yield loss is approximately equal to the percentage of infected ears and ranges from 10 to 40%. In contrast to some other smuts, quality of the harvested grain is not affected. Losses in most developed countries are minor, owing to effective disease control practices.

Control

Cultural practices including the use of certified, pathogen-free seed are very effective in controlling loose smut since infected seed is the only source of inoculum.

Fungicide seed treatment is the principal method of control. The systemic fungicide carboxin is very effective in eradicating the pathogen from infected seed, thus preventing disease development. Other, more recently developed, systemic fungicides are also effective seed treatments for control of loose smut. Before the availability of effective fungicides, seed treatment consisted of hot- or cold-water dips. Although effective, these treatments reduced germination by about 10%.

Disease-resistant cultivars are effective in controlling disease and have been durable, even though races of the pathogen exist. The effectiveness of fungicide seed treatments has lessened the role of resistant cultivars in most areas.

Other

Alternaria, Bipolaris, Cladosporium species

Diseases: Black Point (Kernel Smudge) and Sooty Mould

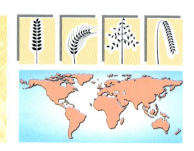

This group of diseases, caused by several generally weak pathogens, can occur on all cereals when prolonged wet weather results in delayed maturation and harvesting. Black point is characterized by brown or black discoloration of the embryo region of the grain. It is generally associated with infection by *Alternaria* and *Bipolaris* species. Sooty mould consists of superficial grey-black fungal growth over the surface of the ear (**27, 28**). Masses of spores may be shed during harvesting (**29**). *Cladosporium* and *Alternaria* are usually considered as the main fungal species responsible for this problem.

27. Sooty mould developing in the foreground of ripe wheat crop.

28. Sooty mould on ripe wheat ear.

29. Sooty mould on ripe wheat ears.

Disease cycle

The fungi responsible for the diseases are ubiquitous facultative parasites with numerous hosts. They can also grow saprophytically on plant debris and are abundant in airspora in temperate regions. Infection of cereals occurs usually during wet weather between flowering and full maturation of the grain. Cereals stressed by other diseases, particularly take-all and eyespot, appear to be especially prone to sooty mould.

Economic importance

Black point is probably more significant than sooty mould in that it can result in a reduction in the bread-making quality of wheat. Significant contamination of grain with black point may result in discoloured flour containing dark particles. Annual losses attributable to the disease were estimated at £3 million in the United Kingdom as samples with over 15% black point affected grain may be rejected by millers for making white flour. In the United States, black point affected kernels are considered as damaged and only 2% and 4% are allowed in Grade No. 1 and 2 wheats, respectively.

Sooty mould is probably in itself not significant, apart from the resultant unpleasant spore clouds produced during harvesting. However, it may indicate the presence of a more significant stem base or root disease in the crop.

Control

Specific measures directed at control of black point and sooty mould are rarely justified. There is some suggestion that wheat varieties can differ in their susceptibility to black point. Limited trials with fungicides to control the disease have been unsuccessful. A range of fungicides are approved for control of sooty mould in the United Kingdom. Most of them consist of mixtures of active ingredients including benzimidazoles, dithiocarbamates, and azoles.

Grain harvested from affected crops should be stored under cool, well-ventilated conditions to reduce the risk of storage mould development.

Claviceps purpurea
Disease: Ergot

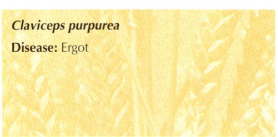

Ergot is a worldwide disease which affects wheat, barley, triticale, and rye. Infection of oats is rare. The disease has been recognized for many centuries, mainly because of the harmful effects of consumption of ergots. The characteristic symptoms of ergot are the presence of horn-shaped, purple–black ergots or sclerotia (10–50 mm long), which appear in mature cereal heads as replacements for individual grains (**30, 31**). However, prior to this, creamy golden droplets of honeydew containing fungal spores may be observed during anthesis in cereals. The honeydew is attractive to insects and may also result in the development of saprophytic moulds on heads.

Disease cycle

Ergots are hard compact masses of fungal tissue which act as resting bodies during inter-crop periods, either in the soil or as contaminants of seed (**32**). After a dormant period at low temperatures, the ergot germinates, producing stalked stromata bearing perithecia which release wind-dispersed ascospores. During humid weather, ascospores alight on cereal florets, germinate, and penetrate the ovary. Infected florets can release asexually produced conidia in honeydew which is dispersed by insects and rain-splashed to other florets. Gradually, the production of honeydew stops and in each contaminated ovary an ergot is produced instead of a normal grain. The disease is favoured by wet, cool weather which prolongs the flowering period and encourages spore germination. Should pollination occur before inoculum reaches the floret, the chance of infection is reduced.

30. Ergots replacing grain on infected wheat ear.

31. Ergots replacing seed in wild grass species.

32. Ergots in grain sample from an infected crop.

Economic importance

The disease results in only small yield losses in cereals (a maximum of 10%). However, ergot contamination of cereals is more significant because of the toxic alkaloids in the ergots. These toxins can cause convulsive or gangrenous ergotism and death in humans and animals. They can also cause abortion.

The disease is more common in rye and triticale than in wheat or barley. Recently it has become of increasing significance to producers of hybrid cereal seed where male-sterile lines may have anthers exposed and unpollinated for longer than would occur normally.

Control

Cultural control involves crop rotation and deep ploughing land to bury ergots to a depth in excess of 70 mm. Elimination of susceptible grass weeds, including blackgrass, may also help reduce disease. Ergots can be removed from grain by gravity separation or sieving. In the United Kingdom, cereal seed certification schemes restrict the number of ergots in grain destined for seed.

Disease resistance in crop varieties is not currently available and chemical control of ergot has been investigated on a limited scale in Europe, but with little success.

Septoria tritici (sexual stage: Mycosphaerella graminicola)

Disease: Septoria Leaf Blotch

See Leaf and Stem Diseases.

Stagonospora nodorum (ex. Septoria nodorum) (sexual stage: Phaeosphaeria nodorum)

Disease: Septoria Leaf Blotch and Glume Blotch

See Leaf and Stem Diseases.

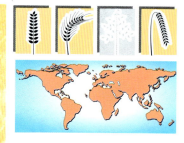

Leaf and Stem Diseases

Blights

Xanthomonas campestris pv. translucens (syn. *X. translucens*)

Diseases: Bacterial Streak, Bacterial Stripe

See Ear and Grain Diseases.

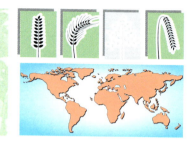

Mildew

***Erysiphe (Blumeria) graminis* f. sp. *tritici* (wheat)**
***Erysiphe (Blumeria) graminis* f. sp. *hordei* (barley)**
***Erysiphe (Blumeria) graminis* f. sp. *avenae* (oats)**
***Erysiphe (Blumeria) graminis* f. sp. *secalis* (rye)**

Disease: Powdery Mildew

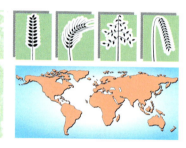

Powdery mildew is probably the commonest disease of cereals, occurring in all areas where crops are grown. Disease symptoms can occur on all aerial plant parts, but are most frequently seen on leaves (**33–37**). Early disease symptoms consist of chlorotic flecks on plant tissue. A white, fluffy mildew pustule soon develops (**38, 39**), which produces masses of powdery spores. Older mildew pustules may assume a grey or brown tinge. The fungus infects only the outer epidermal plant layers and so pustules can be scraped off the leaves relatively easily. Occasionally, mild chlorosis can be seen in affected tissues, particularly at the beginning of natural leaf senescence, but the pathogen does not usually kill its host. Towards the end of the season, brown–black sexually produced spore cases (cleistothecia) may be found embedded in mildew pustules (**40**).

33. Powdery mildew developing on the lower leaves of a maturing wheat crop.

34. Severe powdery mildew infection and distorted wheat ears.

35. Powdery mildew on leaves of a maturing oat crop.

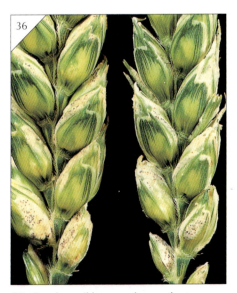

36. Powdery mildew pustules on wheat ears.

37. Powdery mildew development on young barley plants.

38. Photomicrograph of powdery mildew pustules developing on a barley leaf.

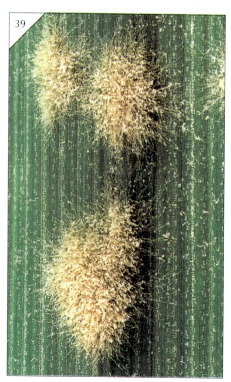

39. Photomicrograph of older powdery mildew pustules on a wheat leaf.

40. Photomicrograph of powdery mildew pustules with cleistothecia.

Erysiphe graminis is an obligate parasite with several specialized forms (f. sp.) There is further specialization within the forma specialis into races or pathotypes that can attack only particular varieties.

Disease cycle

In Europe, the fungus survives the winter mainly as dormant mycelium in host tissue (volunteer plants and over-wintering crops). Cleistothecia are probably unimportant in this respect although they may enable the fungus to survive in debris in the absence of the crop for several weeks. Infections of autumn-sown crops during autumn arise from wind-dispersed conidia (**41, 42**), although ascospores produced from cleistothecia in plant debris may also contribute to inoculum. Conidia germinate over a wide temperature range (5–30°C), but temperatures of 15–20°C accompanied by a few hours of high relative

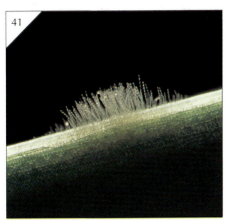

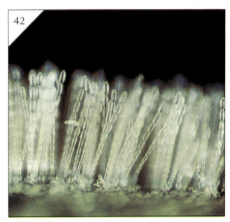

41. Photomicrograph of a mildew pustule on a barley leaf.

42. Photomicrograph of a mildew pustule showing chains of conidia.

humidity (over 90%) are optimal for germination. Free water tends to inhibit spore germination and some spores will germinate at only 80–85% relative humidity. Under optimal conditions the latent period is 7 days; mildew epidemics tend to occur during warm weather with alternating dry and wet periods accompanied by breezes to disperse spores. Disease development is inhibited at temperatures over 25°C. Powdery mildew tends to be severe in lush, over-fertilized, early sown winter cereal crops.

Economic importance

Powdery mildew was the second most common disease of winter wheat and the most widespread and severe disease of spring and winter barley in national surveys carried out over several years in England and Wales. Yield losses in the field in susceptible crops have been estimated at up to 25% in the United States and 20% in the United Kingdom.

Control

Cultural control of powdery mildew involves the eradication of volunteer cereals, which can harbour inoculum over winter, together with disposal of crop debris, which may be infected with cleistothecia. Avoiding very early sowing and excess nitrogen fertilizer applications will also help to reduce disease.

Disease resistance is important in all cereals. Much of the resistance, however, is major gene and as such can break down as a result of a shift in virulence in pathogen populations. Field-by-field diversification of varieties or mixtures of appropriate varieties within fields may slow down disease epidemics.

Chemical control of powdery mildew, particularly in barley and wheat crops, is widely practised. Initially, systemic seed treatments containing azole fungicides may help to reduce disease for the first few weeks after crop emergence. Foliar-applied fungicides can be used on crops at the start of disease epidemics. Morpholine-based fungicides are most frequently used in mildew control. Azoles were used effectively for many years, but now there is tolerance, particularly in populations of *E. graminis* f. sp. *hordei,* to azoles and their use against existing powdery mildew infestation is not widespread.

Mosaics/Yellows

Barley stripe mosaic virus (BSMV)
Disease: Barley Stripe Mosaic

Barley stripe mosaic is also known as barley false stripe, barley mosaic, and oat stripe mosaic. The principal economic host is barley, although wheat, rye, and several grasses are also infected. The disease was first described in Wisconsin, United States in 1910, but is distributed worldwide. BSMV is a member of the Hordeivirus group. Virus particles are rigid rods, of size 20×100 nm and 20×150 nm.

Symptoms of barley stripe range from latent (i.e. infected plants are symptomless) to lethal necrosis wherein infected plants are killed. Other symptoms include yellow to white mottling (**43, 44**), spotting, and streaking of young leaves. In some cases, leaf blades may be completely white. Older leaves often have brown, necrotic, longitudinal, or V-shaped stripes along with mottling or mosaic symptoms (**45, 46**). Infected plants

43. Early symptoms of barley stripe mosaic on barley (courtesy of Professor Thomas W. Carroll, Montana State University).

44. Symptoms of barley stripe mosaic on barley (courtesy of Professor Thomas W. Carroll, Montana State University)

45, 46. Symptoms of barley stripe mosaic in older barley plants (courtesy of Professor Thomas W. Carroll, Montana State University)

are moderately to severely stunted, have poorly developed ears, and have varying degrees of pollen sterility.

Disease cycle

The primary inoculum for barley stripe mosaic is infected seed. Virus replication begins in seedlings during germination and the virus spreads systemically throughout the plant. The virus is mechanically (sap) transmissible and plant-to-plant contact is responsible for spread within fields during the growing season. Both pollen and ovules of infected plants contain the virus, although infected pollen has poor viability and transmission to seed is primarily via the ovule. BSMV remains infective in seed for several years.

Symptom expression is optimal in the range 22–30°C with relatively high light intensity. Inadequate light intensity and temperatures less than 20°C inhibit symptom development. Maximum seed transmission occurs at 20–24°C.

Economic importance

Barley stripe mosaic virus is one of the few seed-transmitted viruses of small grain cereals, with no known insect vectors. Seed transmission rates up to 100% may occur depending upon the cultivar, growth stage when infection occurs, virus strain, and environmental conditions. Yield loss likewise depends on these variables, but may range up to 40% and is the result of fewer kernels per ear and reduced kernel size.

Control

The use of virus-free, certified seed is the primary control for barley stripe mosaic. Control of volunteer (self-sown) plants within fields mechanically, chemically, or with crop rotation is also useful in disease control.

Resistance to seed transmission and mechanical inoculation is available, although it is not widely used owing to the effectiveness of seed certification. In addition, numerous strains of the virus exist, complicating the development of virus-resistant cultivars.

Barley yellow dwarf, covering several viruses (RPV, MAV, PAV, SGV, RMV)

Disease: Barley Yellow Dwarf

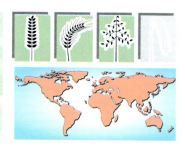

BYDV has a wide host plant range including wheat, barley, oats, and many cultivated and wild grasses. It occurs in all parts of the world where cereals are cultivated and is considered a major threat, particularly to barley crops. BYDV is a member of the Luteovirus group.

Particles are spherical, with diameters in the range 25–30 nm. Symptoms of BYDV are highly variable within one cereal crop and differ from crop to crop. However, dwarfing is a common feature of the disease in all crops. In barley, a bright golden yellowing appears

47. Barley crop with foci of BYDV infection.

48. Barley plants with typical symptoms of BYDV infection.

49. Wheat crop with foci of BYDV infection.

50. Symptoms of BYDV on the flag leaves of an infected wheat crop.

on affected leaves which usually begins from the leaf tip or margins and progresses towards the base (**47, 48**). The tissue nearest to the midrib may remain green longer than the rest. Wheat can show similar yellow discolorations, but this is often accompanied by a red to purple colour (**49, 50**). In oats, a distinctive red-purple discoloration of leaves usually occurs (**51, 52**). BYDV is transmitted by aphids. In Europe, RPV is transmitted specifically by *Rhopalosiphum padi* (**53, 54**), MAV by *Sitobion avenae* and/or *Metopolophium dirhodum*, and PAV non-specifically by all three cereal aphid species. In North America a further two strains have been classified: these are SGV transmitted by *Schizaphis graminum*

51. BYDV causing typical 'redleaf' symptoms on young oat crop.

52. Typical symptoms of BYDV on oat crop flag leaves.

and RMV transmitted by *Rhopalosiphum maidis*.

Disease cycle

During inter-crop periods, BYDV persists in over-wintering cereals, volunteer cereals, grasses, and its aphid vectors. Spread of BYDV

53. Bird cherry oat aphids (*Rhopalosiphum padi*), a vector of BYDV.

to cereal crops is entirely dependent upon migration of its aphid vectors. During autumn (in early sown winter crops) and spring, viruliferous aphids over-wintering on cereals or grasses migrate to young cereal crops to feed and the virus is transmitted. Non-infected aphids can then acquire the virus by feeding on infected plants for as little as 30 minutes, but generally a longer feeding period is required (around 24 hours). BYDV does not multiply in its vector, but a latent period of a few days is needed before it can transmit the virus. Aphids can remain infective for several weeks. Disease symptoms usually occur about 2 weeks after infection and symptom expression is favoured by bright, sunny weather. The disease is most severe during moist, relatively cool (10–18°C) conditions, which favour aphid multiplication, secondary infection, and migration. During late summer, aphids migrate to early sown winter cereals or grass. The migration usually stops during autumn.

Economic importance

BYDV is considered to be the most important viral disease of small grain cereals in the world. Losses of 40% are not uncommon in commercial barley crops, and in wheat BYDV is frequently estimated to reduce yields by up to 25%.

54. Bird cherry oat aphids (*Rhopalosiphum padi*), a vector of BYDV.

In oats, inoculation of field plots with BYDV resulted in a total yield reduction of 33%. Early infection usually results in lower yields.

Control

Cultural control of BYDV involves late sowing of crops and elimination of grass weeds and self-sown cereals. Crops that follow permanent pasture or grass are at risk and attempts should be made to kill residual grass before cultivation. A 2-week period should then be allowed before sowing the cereal crop.

Disease resistance to BYDV is actively being sought in cereal varieties. However, to date, varieties have proved to be tolerant, but not immune, to infection.

Chemical control of the aphid vectors is practised frequently in Europe. In the United Kingdom, disease epidemics are forecast by monitoring local populations of migrating viruliferous aphids. Application of insecticides containing active ingredients such as pyrethroid and organophosphorus should be related to such forecasts.

Barley yellow mosaic virus (BaYMV)

Barley mild mosaic virus (BaMMV)

Disease: Barley Yellow Mosaic

BaYMV is a disease almost exclusively of autumn-sown barley (**55**), and is now quite common in Europe. It was first identified in Japan in 1940, and was reported in West Germany in 1978 and in France and the United Kingdom during 1980. Early symptoms of the disease seen from late December to March are chlorotic or pale green spots and streaks along the leaf veins (**56, 57**). In some varieties the streaks become necrotic. Plants may be stunted and commonly have a spiky appearance as leaf edges roll inward. It is common for early symptoms to disappear during summer, but affected mature plants are often stunted and contain fewer fertile tillers. BaYMV and BaMMV are members of the Potyviridae group. The particles are long, flexuous rods approximately 12 nm wide and range in length from 275 nm to 550 nm.

Disease cycle

BaYMV relies on its fungal vector *Polymyxa graminis* for its transmission and survival in the absence of a suitable host. BaMMV is also

55. BaYMV foci of infection in a young barley crop.

56. BaYMV symptoms on leaves from an infected barley plant.

57. BaYMV symptoms on sampled leaves from an infected barley crop.

transmitted by the same vector and results in similar symptoms. *P. graminis* forms resting spores that remain viable in the soil for many years and there is evidence that the virus can remain viable in such spores. Biflagellate zoospores are released from resting spores, which swim in soil moisture and infect barley roots via root hairs or epidermal cells. If the zoospores are virus infected and the variety is susceptible, then the plant may become infected. Plasmodia are formed in affected barley roots and these develop into zoosporangia from which secondary zoospores are produced. Secondary infections can result from these zoospores if they are virus infected. Later in the season grape-like clusters of resting spores form from plasmodia that are released into the soil or persist on stubble and crop debris. The disease can be spread by cultivation of contaminated soil and frequently follows cultivation lines. Contaminated soil can also be spread on machinery and boots. The disease may be more of a problem in poorly drained waterlogged soils and during particularly cold winters.

Economic importance

Yield losses of between 10 and 90% have been reported in winter barley crops, depending upon variety, climate, soil type, and inoculum potential of soil. In the absence of resistant varieties the disease can make winter barley cultivation uneconomic.

Control

Control of the disease is almost entirely reliant on the use of resistant varieties. Several BaYMV-resistant winter barley varieties with satisfactory agronomic traits are commercially available in Europe. However, it is believed that the basis of such resistance is a single recessive gene, and the appearance of resistance-breaking strains of BaYMV has occurred in several European countries, including Germany and the United Kingdom.

Wheat soil-borne mosaic virus (WSBMV)
Disease: Wheat Soil-Borne Mosaic

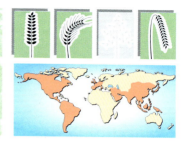

Wheat soil-borne mosaic was first described in Illinois, United States, in 1919 as wheat rosette, a disease causing a mosaic and severe stunting in some cultivars of winter wheat. The disease has since been reported throughout the hard and soft red winter wheat-growing area of the United States, Canada, Japan, China, Egypt, Brazil, Argentina, France, and Italy. Wheat, barley, and rye are susceptible to infection, but the disease is primarily a problem of winter wheat.

WSBMV are members of the Furovirus group, and are rigid rods of size $20 \times 280–300$ nm and $20 \times 140–160$ nm. Several strains of WSBMV exist, resulting in symptoms that range from mild to severe mosaic and moderate to severe stunting. Symptoms appear in the spring on young leaves as a yellowish-green or bluish-green mosaic, mottle, and/or streaking (**58, 59**), and often diminish in severity as temperature increases in the spring. Extreme stunting occurs in certain cultivars and is referred to as wheat rosette. Infected plants are frequently localized within fields along waterways (**60**), in low, wet areas, and around old building sites. Symptoms of wheat soil-borne mosaic may be confused with nutrient deficiencies, herbicide damage, and other diseases.

Disease cycle
WSBMV survives in soil in association with resting spores of *Polymyxa graminis*, a root-infecting fungus. Resting spores germinate by forming zoospores that swim in wet soil and infect root hairs. *P. graminis* subsequently colonizes root cortical tissues, producing more zoospores and resting spores in the process. WSBMV is apparently borne within zoospores and infects plants following penetration of the fungus. WSBMV is also mechanically transmissible, but the importance of this mode of transmission under field conditions is unknown. Infections occurring in the autumn are most important for disease

58. Yellow and green islands on wheat leaf infected with WSBMV (courtesy of Dr Robert L. Bowden, Kansas State University).

59. Symptoms of wheat soil-borne mosaic on young wheat plants (courtesy of Dr Robert L. Bowden, Kansas State University).

60. Symptoms of wheat soil-borne mosaic in a field of wheat (courtesy of Dr Robert L. Bowden, Kansas State University).

development since the disease has a long incubation period. Although infection may occur in the spring, there is not sufficient time for disease to develop.

Soil temperatures from 12 to 15°C and high soil moisture are most favourable for infection. Short daylengths and temperatures less than 17°C are favourable for symptom development. Symptoms are very mild above 25°C.

Economic importance

Crop damage is related to cultivar susceptibility, soil moisture and temperature in the autumn, and date of sowing. In general, earlier sowing in the autumn results in greater disease. Loss in grain yield results from fewer tillers per plant plant, decreased kernel weight, and test weight (specific weight). Losses due to rosette can be complete; however, this extreme form of the disease occurs in relatively few cultivars. Average losses due to mosaic range from insignificant to 20%.

Control

Cultural practices such as delaying sowing of winter wheat in the autumn reduce disease, probably owing to reduced activity of *P. graminis* at lower temperature. Crop rotation with non-hosts is of limited value since resting spores and the virus remain viable in soil for long periods of time.

Disease resistance is the primary control for wheat soil-borne mosaic. It was recognized very soon after discovery of the disease that lines immune to rosette and resistant to mosaic could be selected from susceptible cultivars.

Chemical control through soil fumigation effectively controls *P. graminis*, but is not cost-effective.

Wheat streak mosaic virus (WSMV)
Disease: Wheat Streak Mosaic

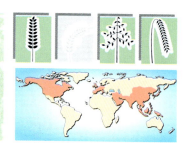

Wheat streak mosaic was described in 1937 from the Central Plains of the United States and is now known to occur in Canada, France, Jordan, Rumania, Yugoslavia, Turkey, and the former Soviet Union. WSMV infects wheat, barley, oats, rye, maize, and several grasses, but the disease is most important on winter wheat. WSMV is a member of the Potyviridae family of viruses. Particles are flexuous rods approximately 12 nm in diameter and 700 nm long. WSMV is transmitted by the wheat curl mite,

Eriophyes tulipae (syn. *Aceria tulipae*), in most parts of the world, and by *E. tosichella* (syn. *A. tosichella*) in the former Yugoslavia. The virus is also mechanically transmissible, but the importance of this mode of transmission under field conditions is unknown.

Distribution of disease within fields depends upon distribution of the wheat curl mite and varies from widespread to localized along field borders. Symptoms appear on infected plants in the spring as temperature

61. Chlorotic streaks in wheat leaf due to wheat streak mosaic virus.

62. Young wheat with symptoms of wheat streak mosaic virus (courtesy of Dr Robert L. Bowden, Kansas State University).

63. Chlorotic streaks in wheat leaves due to wheat streak mosaic virus.

increases and range from mild, light-green or yellow streaking and mottling to severe, yellow spotting, streaking, or mottling (**61–63**). Mosaic symptoms become necrotic on older plants and infected stems are stunted with wholly or partially sterile ears. Mites feed on young, succulent tissues such as expanding leaves, causing them to roll. Subsequent leaves are often trapped in the rolled leaves and are unable to expand.

Disease cycle

Survival of WSMV and the wheat curl mite through the summer months is the most important factor determining the incidence and severity of wheat streak mosaic in the subsequent winter wheat crop. Late-maturing spring wheat and/or volunteer (self-sown) wheat plants are the most important over-summering hosts for both the virus and vector. All growth stages of the mite, except eggs and older adults, are capable of acquiring WSMV from infected plants. Mites reproduce prolifically during warm weather and crawl up to leaf tips where they are picked up and carried up to several kilometres by wind currents to other host plants. Primary inoculum is in the form of viruliferous mites that move from infected host plants to newly emerged winter wheat seedlings. Viruliferous mites also disseminate the virus within the crop and therefore, serve as the source of secondary inoculum. Mite reproduction and virus replication occur throughout the growing season during favourable environmental conditions.

Reproduction of mites and replication of WSMV are favoured by temperatures of 24–27°C. The wheat curl mite can complete a single generation in 7–10 days under such conditions. Rain and/or hail storms in late summer promote growth of volunteer plants that can serve as reservoirs for the mite and pathogen.

Economic importance

Wheat streak mosaic can be devastating and cause complete loss of a crop, depending upon virus strain, time of infection, cultivar, and environmental conditions. Yield loss results from sterility of ears (fewer kernels per ear), reduced test weight, and premature death of infected stems.

Control

Sanitation practices that eliminate over-summering hosts of the mite and virus before planting winter wheat in the autumn effectively controls wheat streak mosaic. Tillage to destroy volunteer plants or other hosts should occur at least 1 week before sowing since mites can survive for several days on detached leaves buried in soil. Delaying sowing in the autumn until over-summering hosts have matured or are no longer green also controls disease.

Resistance to virus replication is present in some wheat relatives (*Agropyron* and *Secale*) and efforts to transfer the resistance to wheat are in progress. Resistance to mite feeding is available and has been used successfully to reduce disease.

Wheat Yellow Mosaic Virus (WYMV)

Diseases: Wheat Yellow Mosaic and Wheat Spindle Streak Mosaic

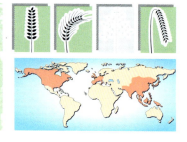

Wheat yellow mosaic and wheat spindle streak mosaic [caused by wheat spindle streak mosaic virus (WSSMV)] were described from Japan and Canada in about 1960 and 1970, respectively. WYMV is a member of the Potyviridae family. Particles are flexuous rods approximately 12 nm in diameter and 275–300 nm and 575-600 nm long. These viruses have many similarities and, although not identical, WSSMV is now considered to be a strain of WYMV. Wheat yellow mosaic occurs on winter wheat grown in areas where cool (less than 17°C) temperatures prevail during much of the growing season, including Canada, the United States, India, France, Japan, and China. The disease is similar to, and may be confused with, wheat soil-borne mosaic.

Symptoms of disease appear in the early spring on winter wheat and increase in severity from the youngest to the oldest leaves. Younger leaves have distinct light-green to yellow, spindle-shaped dashes and short streaks oriented parallel to leaf venation (**64**). Severity of the mosaic symptoms may increase on older

64. Early season symptoms of wheat spindle streak mosaic in wheat (courtesy of Dr Robert L. Bowden, Kansas State University).

65. Symptoms of wheat spindle streak mosaic on wheat leaves (courtesy of Dr Robert L. Bowden, Kansas State University).

66. Symptoms of wheat spindle streak mosaic in a field of wheat (courtesy of Dr Robert L. Bowden, Kansas State University).

leaves such that the spindle-shaped lesions become necrotic in the centre and have a light-brown colour (**65**). Infected plants produce fewer tillers and are slightly stunted. The distribution of disease within fields tends to be less patchy and more uniform than with wheat soil-borne mosaic (**66**).

Disease cycle

The disease cycle for wheat yellow mosaic is very similar to wheat soil-borne mosaic and barley yellow mosaic. Survival of the virus is in association with resting spores of *Polymyxa graminis*. Germination of resting spores results in motile zoospores that swim in wet soil and infect root hairs. Whether WYMV is located on the surface of the zoospore or contained within is unknown, but infection presumably occurs following penetration of the root hair by the fungus. Like wheat soil-borne mosaic virus, WYMV is mechanically transmissible, but the importance of this mode of transmission under field conditions is unknown. Infections in the autumn are most important for disease development.

WYMV may have the lowest temperature requirement of the cereal viruses. The optimum temperature for infection is 5–13°C and prolonged (60 days) temperatures from 5–15°C with relatively low light intensity favour symptom development. Temperatures over 20°C suppress disease development.

Economic importance

Disease severity varies with cultivar susceptibility, date of sowing, and the frequency of wheat production within a field. Yield losses are the result of fewer tillers per plant and fewer kernels per ear and range up to 64%.

Control

Cultural practices reduce disease severity, including delayed sowing in the autumn. Long crop rotations are associated with reduced disease severity; however, this is not practical in most areas.

Disease resistance is available and has been used successfully to control disease. Disinfestation of soil with heat (55°C for 30 minutes) or fumigation effectively controls *P. graminis*, but is not cost-effective in most areas.

Rusts

Puccinia coronata f. sp. *avenae*

Disease: Crown Rust

Crown rust is a disease that predominantly affects oats. Specific races or pathotypes of the pathogen exist that attack particular varieties of oats. In recent years a distinct form of the pathogen has been identified in the United States that can attack barley. It is considered to be the most important disease of oats on a world-wide basis and is particularly important in temperate humid regions. Symptoms of crown rust on oats consist of bright orange elongated pustules 1–5 mm long, which occur on all aerial parts of the plant, but primarily on leaves (**67–69**). Dense patches of pustules may occur. Later in the season black telia may form in lines on leaf sheaths (**70**).

67. Crown rust infection on oat crop.

68. Crown rust pustules (uredinia) on infected oat leaf.

69. crown rust pustules (uredinia) on infected oat leaf.

70. Crown rust damage and pustules (telia) in infected oat crop.

Disease cycle

The pathogen survives the inter-crop period as uredinia, telia, and dormant mycelium on host plant debris, self-sown oats and over-wintering crops. Early infections can arise from these sources, but spores (aeciospores) produced on the alternate hosts *Rhamnus cathartica* and *Frangula alnus* may also initiate the disease. The alternate host can become re-infected by basidiospores produced from germinating teliospores. Such teliospores have a flattened apex with a crown of 5–8 projections (hence crown rust). Further aeciospores may then be produced on the alternate host, which can re-infect oats later in the season. Disease epidemics, however, usually arise from asexual, wind-dispersed urediniospores, which are produced optimally during warm (20–25°C) moist (free water) weather. The latent period is 7 days under such ideal conditions for disease development.

Economic importance

Yield losses in the range 10–20% have been reported in some areas of the United States as a result of a crown rust epidemic. In a recent investigation of the effect of disease in two spring oat cultivars, it was estimated that for each 1% increase in crown rust severity, average yield loss was over 50 kg/ha.

Control

Cultural control methods include the eradication of volunteer cereals, which can harbour inoculum over winter, together with the disposal of crop debris. Avoiding very early sowing and excess nitrogen fertilizer applications also helps to reduce the disease. In addition, elimination of known alternate hosts in the vicinity of crops is a sensible precaution.

Disease resistance has been incorporated into commercially available oat varieties, but there have been problems with resistance breakdown. 'Slow-rusting' varieties offering more durable resistance are being sought.

Chemical control using foliar-applied fungicides containing azole active compounds are available in many countries, but their economic use could be justified only on high-value crops (e.g. seed).

Puccinia graminis f. sp. *tritici* (wheat, barley)
Puccinia graminis f. sp. *avenae* (oats)
Puccinia graminis f. sp. *secalis* (rye)

Disease: Black Stem Rust

Black stem rust is one of the most widespread and important diseases of cereals worldwide. Many historical accounts describe epidemics brought about by black stem rust. The true nature of the disease was not understood until 1865, when it was demonstrated that *P. graminis* requires both a small grain cereal and the barberry (primarily *Berberis vulgaris*, but also *B. canadensis*) or *Mahonia* sp. to complete its life cycle. In addition to the small grain, *P. graminis* can infect many other wild and cultivated grasses.

Puccinia graminis is an obligate parasite with many specialized forms, each of which is capable of infecting one or a few host species. For example, *P. graminis* f. sp. *tritici* can infect wheat, barley, and triticale, but is unable to infect oats and most varieties of rye. Likewise, *P. graminis* f. sp. *avenae* can infect oats, but not wheat, barley, or rye. In addition, races of the pathogen specialized to cultivars within a host species occur in formae speciales.

Symptoms of black stem rust are not very prominent and include small, raised, yellow-orange lesions on leaves, petioles, and blossoms of infected barberry plants and yellow or brown flecks on the small grain host. Signs of the pathogen are most apparent and provide the diagnostic structures needed to differentiate black stem rust from brown rust, which it resembles. Diamond-shaped pustules (less than 1 cm long) appear first on the leaves, but later may occur anywhere on the plant (**71, 72**) and can coalesce to form large

71. Uredinia of black stem rust on wheat stems.

72. Uredinia of black stem rust on a wheat ear.

73, 74. Uredinia of black stem rust on wheat.

lesions on the plant surface. Pustules (sori) contain urediniospores (repeating spores or summer spores) and give infected plants a rust-red colour (**73**). Later, as plants approach maturity, black teliospores are produced in the pustules, hence the name black stem rust. Another feature that distinguishes black stem rust from brown rust is the appearance of the pustule: uredinial pustules of black stem rust have large, light-coloured flakes of ruptured tissue surrounding them (**74**), a feature that is absent or greatly reduced with brown rust. Signs on barberry include groups of salmon-pink, tubular cup-like aecia (cluster cups) on leaves, petioles, and fruits.

Disease cycle

Puccinia graminis is a macrocyclic, heteroecious rust (requires both a cereal grain and barberry). In most temperate areas, *P. graminis* survives as teliospores in crop debris from the previous infected crop. However, the pathogen is capable of surviving as mycelium in live plants in areas with mild winter temperatures. Teliospores germinate in infested residue in the spring, producing basidiospores that are wind disseminated for up to a quarter of a mile to barberry leaves, which they penetrate directly through the cuticle. A spermagonium forms on the upper leaf surface at the penetration point and the aecia form soon thereafter on the lower leaf surface, beneath spermagonia.

Aeciospores are wind disseminated to the cereal plant, which the fungus penetrates indirectly via stomata. The uredinia form in the vicinity of the penetration point and rupture the plant epidermis as the urediniospores are formed. Urediniospores are wind disseminated (up to several thousand miles) and act as secondary inoculum, infecting other cereal grains. When environmental conditions are favourable for disease, only 7–10 days are required from inoculation to production of urediniospores; thus, many disease cycles occur each growing season. Teliospores are formed as the host senesces and/or night temperatures become cooler.

Environmental conditions have a significant impact on disease development. Low temperatures or freezing and thawing during winter are favourable for breaking dormancy of teliospores. Infection of both the cereal and barberry plants requires several hours of free moisture on the plant surface and relatively warm conditions. Black stem rust develops optimally near 24°C and slows greatly below 15°C. For these reasons, black stem rust is more severe in summer rainfall areas (continental climate) than winter rainfall areas (maritime climate).

Economic importance

Grain yield and quality can be reduced significantly when black stem rust is severe. Infected plants are tillerless, produce fewer roots, are predisposed to winter injury, and have reduced photosynthetic leaf area and increased evaporative water loss. The degree of loss depends upon prevailing weather conditions, time of infection, and susceptibility of the cultivar being grown, but may be complete when disease is severe.

Control

Disease resistance has provided the most effective method of control for black stem rust. Extensive effort has been placed on development of cultivars with race-specific resistance. Unfortunately, *P. graminis* is a highly variable pathogen that has often cir-

cumvented this type of resistance within a few years of the release of a new cultivar. As a consequence, several hundred races of the pathogen have been identified. Strategies involving the incorporation of multiple race-specific disease-resistance genes in the same cultivar (gene pyramiding) and the use of different resistance genes in adjacent growing regions (gene deployment, varietal diversification) have been used to increase the durability of resistant cultivars.

Legislative control involving eradication of the common barberry has been practised since about 1660, when France passed the first barberry eradication laws. Eliminating the barberry to disrupt the disease cycle has had mixed success in different production regions, depending upon the source of inoculum. Barberry eradication is effective in areas where the barberry is the only source of inoculum for the cereal crop, but is not effective in areas where *P. graminis* can survive the winter as mycelium in infected plants or where inoculum from other growing areas is deposited by wind.

Chemical control with both protective and systemic fungicides is effective in controlling black stem rust. Systemic fungicides, such as the azoles, are preferred because their residual activity is longer under weather conditions that favour disease development. Azoles also form part of some systemic seed treatments available to suppress the disease in seedling tissue early in the season. Protective fungicides are also effective, but are easily washed from the plant by rain and must be applied repeatedly during the growing season.

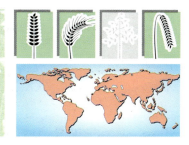

Puccinia recondita f. sp. *tritici* (wheat mainly, also barley, rye); *Puccinia hordei* (barley)

Disease: Brown Rust; Leaf Rust

Brown rust is an important disease of both wheat and barley and it occurs fairly regularly wherever the cereals are grown. It is most important in temperate climates. Typical symptoms of the disease consist of orange–brown pustules on leaves scattered at random (**75–78**). Individual pustules may be slightly larger than those of yellow rust and they are often surrounded by a chlorotic halo (**79**). Under high disease pressure the cereal head may become affected (**80**). Towards the end of the growing season, grey–black telia may form in older diseased areas (**81**).

76. Brown rust infection on wheat flag leaves.

75. Brown rust infection on rye crop.

77. Brown rust (uredinia) on a wheat leaf.

78. Early infection of brown rust on young barley (powdery mildew is also present).

79. Brown rust pustules (uredinia) on a barley leaf.

80. Wheat crop in ear severely affected by brown rust.

81. Brown rust pustules (uredinia and telia) on barley leaves and sheath.

Disease cycle

As with yellow rust, the sexual stage of the disease is probably unimportant in most barley-growing areas although alternate hosts have been identified in Israel and Greece. The arable weed, Star of Bethlehem (*Ornithogalum umbelatum*) was recently suggested to be responsible for outbreaks of barley leaf rust (*P. hordei*) in South Australia. Asexual uredin-iospores are important in disease epidemics (**82**). The fungus survives inter-crop periods as dormant mycelium or urediniospores on

82. Photomicrograph of urediniospore pustules of brown rust on a wheat leaf.

volunteer or over-wintering barley crops. Temperatures in the range 15–22°C, together with periods of 100% humidity, are optimal for disease development. Dry, windy days to disperse spores together with cool nights with dew also favour disease. Under such conditions, the latent period may be as little as 6 days. Sporulation is reduced at temperatures over 25°C. The disease is favoured by warmer conditions than those for yellow rust and often occurs at mid to late summer on cereals in temperate climates.

Economic importance

In national surveys of winter wheat diseases in the United Kingdom (1976–1988), brown rust was more frequently observed than yellow rust in every year, but the average percentage area of flag leaf affected by the disease did not exceed 0.5%. In the United States, losses in yield from leaf rust in winter wheat were estimated at 0.9, 2.2, 3.3, and 4.8% in 1989, 1990, 1991, and 1992, respectively. In national UK surveys of spring barley (1976, 1980) brown rust was again more frequently observed than yellow rust with a maximum average of 1.2% flag leaf area affected by the disease in 1979. In similar surveys of winter barley (1981–1991), brown rust was considered as one of the major foliar diseases, with epidemics in 1989 and 1990 when average flag leaf areas affected by the disease were 4.6 and 3.0%, respectively. It was estimated that annual losses due to brown rust on barley averaged 1.2% of the United Kingdom national yield over the survey period. In the United States, it was estimated that an average grain yield loss of 0.42% occurred for each 1% increment of leaf rust severity on the upper two leaves at the early dough stage of plant development.

Control

Cultural control methods include the eradication of volunteer cereals, which can harbour inoculum over winter, together with the disposal of crop debris. Avoiding very early sowing and excess nitrogen fertilizer applications also helps to reduce the disease. The deployment of genetic resistance in varieties and chemical control methods are also similar to those used for yellow rust.

***Puccinia striiformis,* f. sp. *tritici* (wheat mainly, also barley). f. sp. *hordei* (barley)**

Disease: Yellow Rust (Stripe Rust)

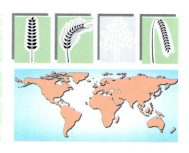

The major economic host of yellow rust is wheat, particularly in cool, maritime regions of the world. In semi-arid areas, only very susceptible varieties are at risk during wet periods. Typical symptoms of yellow rust on mature wheat leaves consist of yellow–orange rust pustules arranged between the veins in stripes – hence the alternative name stripe rust (**83–86**). On young leaves such pustules are scattered at random and may be difficult to distinguish from those of brown rust (leaf rust) caused by *Puccinia recondita* on wheat

or *P. hordei* on barley. Later in the season, yellow rust may occur on cereal heads, resulting in the formation of masses of spores lodged between the glume and the lemma. At the end of the season, black telia may form in necrotic tissue patches killed by the yellow rust pustules.

As with many obligate parasites, e.g. mildew, *P. striiformis* is further differentiated within the f. sp. sub-species such that specific races or pathotypes exist that can infect only particular varieties.

83. Yellow rust infection on a wheat crop.

84. Yellow rust on the flag leaves of a wheat crop.

85. A wheat leaf with advanced yellow rust infection.

86. Stripe development of yellow rust pustules on a wheat leaf.

Disease cycle

The telia that sometimes form in lesions at the end of the season can germinate and form basidiospores, but no alternate host has been found and this sexual phase of the disease cycle is therefore considered unimportant. The fungus survives the inter-crop period mainly as dormant mycelium or uredinia on volunteer cereals. The fungus can survive freezing temperatures, but it may be killed if temperatures fall below –5°C. Disease epidemics occur in warmer conditions (10–15°C), and during periods of high relative humidity. Uredinio-spores are produced (**87, 88**) and wind dispersed locally and over long distances. The latent period of the disease may be as little as 7 days under such conditions. It is common for the disease to start as a focus of infection ('hot spot') in a crop and then to spread initially in relation to the

87. Photomicrograph of yellow rust pustules (uredinia) on a wheat leaf.

88. Photomicrograph of yellow rust pustules (uredinia) on a wheat leaf.

prevailing wind. Warmer conditions (above 20°C) tend to inhibit the disease and thus the problem occurs mainly during spring and early summer in Europe.

Economic importance

In national surveys of winter wheat diseases (1976–1988), it was demonstrated that epidemics of yellow rust occurred only in 3 years (1978, 1981, 1988) in England and Wales. However in susceptible varieties during epidemic years, yield losses of up to 40% have been reported.

Control

Cultural control methods include the eradication of volunteer cereals, which can harbour inoculum over winter, together with the disposal of crop debris. Avoiding very early sowing and excessive nitrogen fertilizer applications also helps to reduce the disease.

Disease resistance is very important in cereal varieties and much emphasis has been placed on breeding yellow-rust-resistant varieties of wheat in many countries throughout the world. Unfortunately, much of the resistance that has so far been incorporated into varieties is major gene and as such has been overcome, sometimes in a spectacular fashion, by adaptation in pathogen populations. The deployment of genetic resistance on farms by the use of mixtures of varieties, either together in a single field or on a field-to-field basis may reduce disease epidemics.

Chemical control is practised in Europe when high-yielding susceptible varieties of wheat are grown. Common active ingredients applied to cereal foliage include azoles and morpholines.

Smuts

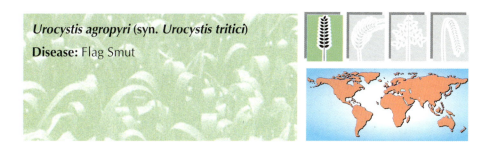

Urocystis agropyri (syn. Urocystis tritici)
Disease: Flag Smut

Flag smut of wheat was first recognized in South Australia in 1868 as 'black rust'. The disease has since been reported from every continent and most countries of the world where wheat is grown. The fungus causing flag smut of wheat was previously named *U. tritici* and considered a distinct species from the pathogen causing flag smut of grasses. However, the morphological similarity and overlapping host ranges of these fungi have resulted in them being considered under the single species *U. agropyri*. Although *U. agropyri* is widespread, flag smut of grasses is more common than flag smut of wheat.

Signs of flag smut are more pronounced than symptoms and include numerous longitudinal pustules between the vascular bundles of leaves, sheaths, awns, and the rachis containing teliospores of the fungus. Infected stems are stunted and distorted, and seldom produce ears (**89, 90**). Immature sori have a white-to-grey colour and become darker as

89. Distortion of wheat flag leaf due to flag smut.

90. Distorted flag leaves and wheat heads with exposed sori of flag smut.

91. Sori and exposed teliospores of stem smut of rye.

92. Distortion of rye heads and flag leaves due to stem smut.

they mature. Eventually the thin covering of the sorus ruptures, exposing the black teliospores. Leaves do not expand fully but remain rolled and twisted. Infected plants may tiller profusely; however, the tillers are slender and spindly. Not all tillers on an infected plant exhibit symptoms. Stem smut of rye, caused by *Urocystis occultata*, resembles flag smut of wheat, however the pathogens causing these diseases are specialized to their respective hosts (**91–93**).

Disease cycle

The disease cycle of flag smut is similar to that of stinking smut. Teliospores are the primary inoculum and may be seed-borne or soilborne. Teliospores are clumped into spore balls containing one to six fertile cells surrounded by several sterile cells. Each fertile cell may germinate and produce one to four basidiospores. Basidiospores germinate by forming a slender hypha that penetrates the coleoptile directly through the epidermis. The hyphae of the fungus grow inter- and intracellularly between vascular bundles of the leaf tissue and other affected plant parts. Individual cells of the hyphae develop into teliospores, which are disseminated by wind

93. Sori and distortion of rye due to stem smut.

following harvest or as a surface contaminant on harvested grain. Teliospores typically survive about 3 years in field soil, although reports exist of survival up to 7 years. Teliospores can also remain viable after passage through farm animals fed infested straw and may represent an additional source of inoculum in some areas.

Infection of wheat is favoured by sowing seed into relatively dry and warm soil. The optimum temperature for infection is 20°C, but infection may occur at as low as 5°C and as high as 28°C. In general, sowing winter wheat early and deeper than 1.5 cm is favourable to the disease.

Economic importance

Loss in grain yield is approximately equal to the percentage of stems infected with flag smut. Complete loss of individual fields has been reported; however, losses in grain yield of 5–20% are more common.

Control

Chemical seed treatments provide very effective control of flag smut. Several protectant fungicides have been used effectively to control disease resulting from seed-borne inoculum. The polychlorobenzenes HCB and PCNB have provided partial control of seed- and soil-borne inoculum. Several systemic fungicides, especially carboxin, but also azole and benzimidazole fungicides, control flag smut when used at appropriate rates.

Cultural practices including shallow (less than 1.5 cm) sowing and sowing when soil moisture and temperature are unfavourable for disease (later in the autumn or earlier in the spring) will limit the incidence of flag smut. A crop rotation of 2–3 years is beneficial in reducing disease, as is deep ploughing to remove the teliospores from the infection court.

Disease resistance is very effective in controlling flag smut, but is not widely used owing to the effectiveness of chemical seed treatments. Races of *U. agropyri* occur; however, resistance has been durable and a proliferation of races has not been observed.

Spots/blotches

Cochliobolus sativus, sexual stage;
Bipolaris sorokiniana (syn.
Helminthosporium sativum, H. sorokini-
anum), asexual stage

Diseases: Common Root Rot, Seedling
Blight, and Spot Blotch

Common root rot and seedling blight occur wherever cereals are grown. All of the small cereal grains and numerous grasses are hosts; however, wheat and barley are the most economically important. These diseases are caused by a complex of unspecialized fungi that can occur together under some conditions. The most important pathogens are *Cochliobolus sativus*, which predominates in the Great Plains of the United States, the prairies of Canada, and the former USSR, and species of Fusarium [see Fusarium seedling blight, foot rot and head (ear) blight]. Although spot blotch, caused by *C. sativus*, occurs wherever wheat and barley are grown, this disease is a significant problem only in areas with warm, humid weather during the growing season.

As indicated by the common name of this disease, all plant parts are susceptible to infection by *C. sativus*, depending upon the prevailing weather conditions. Symptoms of seedling blight include brown, elliptical lesions that progress inward and upward from near the base of the coleoptile below the soil surface. Plants may die before emergence, but usually die after emergence. Common root rot is characterized by dark-brown to black, necrotic lesions on roots, subcrown internodes, and stem bases (**94**). Discoloration of the subcrown internode is characteristic of infection by *C. sativus*. Lesions often coalesce to form large areas of necrotic tissue in the crown. Infected plants are stunted and tillerless, and stems with severe disease may die prematurely, resulting in whiteheads. Symptoms

of spot blotch include uniformly dark-brown, round to oblong lesions on leaves (**95, 96**). Lesions can coalesce, resulting in large areas of infected leaf tissue that dry out. Dark, olive-coloured mats of fungal spores (conidia) form on the lesions during warm and humid weather. Infection of spikelets results in dark-brown, elliptical lesions with light-brown centres on the lemma, palea, and kernels (see black point, kernel smudge, and sooty mould).

Disease cycle

Cochliobolus sativus is an aggressive saprophyte in soil that colonizes and sporulates profusely on infested host debris. The sexual stage is not

94. Severe symptoms of spot blotch on barley leaves (courtesy of Dr R.G. Rees, Queensland Wheat Research Institute).

95. Foliar symptoms (spotting) on barley with spot blotch (courtesy of Dr R.G. Rees, Queensland Wheat Research Institute).

96. Discoloration of barley head due to spot blotch (courtesy of Dr R.G. Rees, Queensland Wheat Research Institute).

involved in the disease cycle in most parts of the world. The primary inoculum for common root rot and seedling blight is mycelium growing from infected seed, conidia on the surface of the kernel, or conidia in soil. The pathogen penetrates plant tissues directly through the epidermis, natural openings, or wounds. Colonization of infected plant parts is followed by sporulation of the fungus. Dissemination of secondary inoculum is not important for continued disease development below ground, but provides inoculum for subsequent crops.

Primary inoculum for spot blotch includes infected seed and conidia from infested plant debris or infected grass hosts. Infection occurs during wet and warm weather and is followed by sporulation of the pathogen on the developing lesions. Production and dissemination of conidia from sporulating lesions continues during favourable weather and is responsible for spread of disease to other plant parts, including ears. Kernels may be infected during any stage of development.

Infection of seedlings and development of common root rot are favoured by relatively warm soils. Disease development occurs in the range 16–40°C, with optimal temperatures of 28–32°C, depending upon cultivar. Moist soils at planting favour infection and colonization by soil-borne inoculum. Disease is most damaging when water stress occurs during kernel development. Wet weather, such as frequent rain showers, and temperatures over 20°C favour development of spot blotch.

Economic importance

Stand density may be reduced owing to seedling blight and plants with common root rot produce fewer tillers per plant and kernels per ear. Estimated losses in grain yield due to common root rot and seedling blight for Canada, Scotland, and Brazil are 15, 10, and 20%, respectively. Losses in grain yield due to spot blotch are in the range 40–85% for Brazil and other non-traditional wheat-growing areas.

Control

Cultural practices, including the sowing of clean, pathogen-free seed, reduce the potential for seedling blight. Delaying seeding of winter cereals to avoid high soil temperatures, and seeding spring cereals as early as possible to avoid warm, humid weather during kernel development, will likewise reduce disease development. Crop rotation provides time for infested residue to decompose, but is only partially effective because the pathogen may survive up to 3 years in the soil.

Disease-resistant cultivars of wheat and barley are available for control of spot blotch and common root rot. Resistance to these diseases is not correlated and may be less effective when environmental conditions are very favourable for disease development.

Fungicide seed treatments, including captan, guazatine plus, iprodione, thiram, and triadimenol, control seed-borne inoculum. Foliar fungicides, including maneb, mancozeb, propiconazole and tebuconazole, are partially effective in controlling spot blotch and reducing inoculum production.

Pyrenophora (*Drechslera*) *teres* f. sp. *teres* - net form; *Pyrenophora* (*Drechslera*) *teres* f. sp. *maculata* - spot form

Disease: Net Blotch

97. Net blotch lesions developing on barley cotyledons.

Net blotch is a disease mainly of autumn-sown barley that can assume two distinct symptomatologies – a net form and a spot form. The latter form is comparatively rare and was classified as a distinct disease only in the late 1960s in Denmark. The disease occurs sporadically in most barley-growing areas of the world. In Europe, net blotch most frequently occurs on leaves of young autumn-sown barley (**97, 98**) and volunteer barley plants during late autumn through to early spring. Longitudinal and transverse dark-brown streaks appear on leaves forming a net-like appearance (**99, 100**). Lesions may also be surrounded by chlorotic tissue (**101**) and severe attacks can result in large dead areas of leaf (**102**). Symptoms of the spot form have been observed in North America, Germany, the United Kingdom,

98. Autumn infection of net blotch on seedling barley plants.

99. Net blotch lesions on barley coming into ear.

Scandinavia, Morocco, and the Middle East. Dark-brown elliptical lesions, again surrounded by chlorotic tissue, appear on leaves.

Disease cycle

The most important primary source of inoculum is probably infested host residues, although the disease can also be seed borne. Wild grass species may also harbour the pathogen. The fungus can reproduce both sexually and asexually. Wind-blown and water-splashed ascospores produced from pseudothecia may initiate primary infections, although it is probable that the large, asexually produced conidia are more important in initiation and spread of disease. Conidia are also dispersed by wind, although some splash dispersal may occur. Most infection occurs during prolonged high humidity (10–30 hours) and temperatures in the range 10–25°C. The pathogen is inhibited by higher temperatures and dry weather; however, under optimal conditions the disease cycle can be completed in under 14 days.

100. Early infection of net blotch on barley cotyledons.

101. Characteristic net blotch lesions on barley leaves.

102. Severe late season infection of net blotch on barley.

Economic importance

The sporadic nature of the disease makes it difficult to assess the overall economic importance of net blotch. In a national survey of United Kingdom spring barley the disease was rare; however, in similar surveys of winter barley, epidemics were recorded in 3 of the 9 years surveyed (1981, 1987, 1988). It was also estimated that net blotch was responsible for between 0.7 and 1.4% losses of national yield. On susceptible varieties in epidemic years, yield losses of over 35% have been reported in the United States. There is also some evidence to show that net blotch can reduce malting quality of barley by affecting the carbohydrate content of grain.

Control

Cultural control methods include the use of high-quality, pathogen-free seed, together with the disposal of crop debris and volunteer plants. Avoiding very early sowing and excess nitrogen fertilizer applications also helps to reduce the disease.

Disease resistance is important in disease control and is available, particularly in European varieties of autumn-sown barley. It is likely that genetic resistance is based on several dominant genes, but it is also possible for the pathogen to overcome such resistance.

Chemical control may be attempted initially by the use of systemic seed treatments containing azole components, which will eradicate seed infection and protect the crop in the early growth stages. Foliar-applied fungicides with active ingredients including azoles, benzimidazoles and morpholines are available, and may be applied if the disease is threatening during late autumn and early spring.

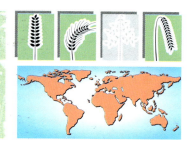

***Pyrenophora tritici-repentis* (syn. *P. trichos-toma*), sexual stage; *Drechslera tritici-repentis* (syn. *Helminthosporium tritici-repentis*), asexual stage**

Disease: Tan Spot (Yellow Leaf Spot)

Tan spot occurs in all of the major cereal growing areas of the world, but is more common and destructive in areas having relatively warm and wet weather during the cereal growing season. Wheat, barley, rye, and numerous grasses are hosts for *Pyrenophora tritici-repentis*, but wheat is the most important.

Tan spot occurs primarily on the leaves and sheaths, even though all above-ground plant tissues are susceptible to infection. Lesions begin as small, tan-to-brown flecks (**103, 104**) that expand into tan-coloured, elliptical lesions (approximately 12 mm long) with dark centres and chlorotic halos (**105**). Lesions may coalesce into large necrotic areas (**106**), causing leaves to wither from the tip. The fungus sporulates profusely on lesions during wet weather, giving them a dark colour.

Disease cycle

P. tritici-repentis survives as a saprophyte on infested host debris between crops. Pseudothecia, approximately 0.2–0.35 mm in diameter, are produced in abundance on straw lying on soil during the autumn and winter. Ascospores are released in the spring during wet weather and serve as primary inoculum. Other sources of primary inoculum include mycelium from infected seed and conidia produced on colonized straw, other grass hosts, and volunteer plants. Conidia produced on primary lesions during wet

103. Tan spot lesions on wheat leaf.

104. Characteristic lesions of tan spot with darkened centre and pale margin.

105. Tan spot lesions on wheat leaves.

106. Severe tan spot infection of a wheat crop.

weather serve as secondary inoculum and are disseminated by wind to other parts of the same plant or other plants. Kernel infection is related to the severity of disease on the flag leaf and occurs when favourable environmental conditions persist into flowering. Kernels may be infected at any time during development, but are most susceptible during the milk stage.

Disease development occurs over a wide range of temperatures, but is optimal from 20 to 28°C, depending upon the cultivar. Frequent rains and wetting of the foliage favour infection and production of secondary inoculum. Minimum leaf wetness duration for disease development depends upon the cultivar and ranges from 6 hours for susceptible cultivars up to 48 hours for resistant cultivars.

Economic importance

Losses in grain yield are primarily the result of reduced kernel size and can reach 50%. Germinability of infected seed is not affected, but vigour of seedlings emerging from infected seed is reduced.

Control

Cultural practices, such as crop rotation with non-hosts and removal or destruction of infested residue, are effective in controlling tan spot. Tan spot is more severe with reduced tillage systems, where crop residues remain on the soil surface than with conventional systems where residues are buried.

Seed treatment with several different systemic fungicides, either individually or in combination, controls seed-borne inoculum. Application of azole, ethylenebisdithiocarbamate (EBDC), or combinations of the two fungicides to the foliage controls disease, but may not be economically feasible.

Disease-resistant cultivars are available and effective in controlling tan spot.

Biological control by the application of antagonistic fungi to colonized residue on soil can reduce inoculum production, but is not effective enough to be used commercially.

Rhynchosporium secalis

Diseases: Barley Leaf Blotch or Scald

Leaf blotch or scald caused by *Rhynchosporium secalis* is mainly a disease of barley, although rye can also be affected. The disease is particularly common in cooler maritime barley growing areas of the world, including many parts of Europe, Scandinavia, North America, Asia, and Australasia. Symptoms usually appear on leaves where initially pale grey-green, water-soaked patches occur (**107**). Within a few days, lesions enlarge and the centre dries out, assuming a pale grey–brown colour and a dark-brown edge develops around the lesion (**108**). Lesions frequently coalesce, resulting in large areas of necrotic tissue (**109**). It is common to find lesions at the junction of the leaf and the stem (leaf axil, **110**). Such lesions can be particularly damaging for the plant as they can cause the leaf to lose its natural erect position, or they may kill the leaf. Occasionally, symptoms can also be seen on leaf sheaths and glumes.

108. Leaf blotch symptoms on barley leaves.

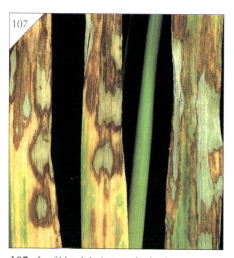

107. Leaf blotch lesions on barley leaves.

109. Severe leaf blotch infection on a maturing barley crop.

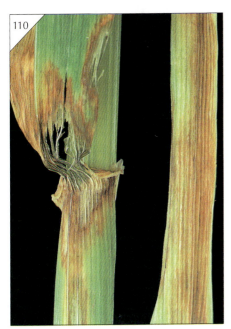

110. Necrosis on a barley leaf with axial characteristics of water-borne leaf blotch.

Disease cycle

Mycelium on cereal debris and seed frequently serves as an initial source of inoculum. Sporulation occurs within 48 hours at 10–18°C during continuous wetness. Primary infections arise mainly from splash dispersal of spores from affected crop debris and the disease cycle may be repeated every 14 days during optimum conditions. Higher temperatures (above 25°C) together with dry weather inhibit sporulation and disease development. During wet weather, spores may be dispersed to ears, which can result in seed infection. The fungus persists for up to 12 months on stubble and debris, but it does not survive well saprophytically in the soil.

Economic importance

In surveys of winter barley diseases in England and Wales (1981–1991), it was estimated that mean annual losses caused by *R. secalis* were 1.0% of national yield. Similar data were obtained in surveys of spring barley. During severe disease epidemics, yield losses in the range 30–40% have been reported in individual crops.

Control

Cultural control involves the use of high-quality, pathogen-free seed, together with disposal of crop debris and volunteer plants. Optimal use of nitrogen fertilizer, together with avoidance of very early sowing, will also help to reduce disease.

Disease resistance is very important in control of scald. Both polygenic and major gene resistance to the disease are available. However, there are still some highly susceptible varieties grown over large areas.

Chemical control of scald is practised widely in Europe, although the practice is relatively rare in North America. In winter barley, a systemic seed treatment including an azole component may be used to reduce seed infection and protect against early attacks of the disease. A range of fungicides, including azoles and benzimidazoles, may be applied to crops to reduce disease, particularly during wet weather. There is resistance in some populations of *R. secalis* to benzimidazole fungicides.

Selenophoma donacis (syn. *Pseudoseptoria donacis*)

Disease: Halo Spot

Halo spot is a relatively minor foliar disease, predominantly of barley, although it can occur on all small grain cereals and many grass species, particularly timothy and cocksfoot. It is usually confined to cool maritime climates. Symptoms consist of scattered oval, pale fawn lesions with a purple-brown border (**111**). Such lesions often occur towards the tips and edges of leaves and they may coalesce, resulting in large areas of necrotic tissue (**112**). Small dark-brown pycnidia can also be pro-duced in lesions, often arranged in rows between veins (**113**).

Disease cycle

Initial sources of inoculum arise from fungus-contaminated seed, debris, and volunteer cereals. During wet weather spores are released from pycnidia and rain dispersed in a similar way to Septoria diseases of cereals. Cool, moist conditions tend to favour the disease, which is most severe in luxuriant over-fertilized crops.

111. Old lesions of halo spot on a barley flag leaf.

112. Small, discreet halo spot lesions on barley leaves.

Upper leaves of barley appear to be more prone to the disease than lower.

Economic importance

The disease was recorded on only a very small percentage of crops in national surveys of winter and spring barley diseases in the United Kingdom over several years. The effect of the disease on yield is considered to be slight.

Control

Few control measures are directed specifically towards the disease because of its perceived lack of economic importance. Fungicides applied both to seed and foliage for other more important diseases may, however, also suppress halo spot.

113. Photomicrograph of characteristic halo spot lesions with pycnidia.

Septoria tritici (sexual stage: Mycosphaerella graminicola)

Diseases: Septoria Leaf Blotch

Septoria leaf blotch is a disease that primarily affects wheat in most cereal growing areas of the world, particularly during wet summers. On oats the disease is caused by *Mycosphaerella graminicola* f. sp. *avenae*.

Septoria tritici lesions are elongate ovals, running parallel to leaf veins (**114, 115**). Grey water-soaked patches appear which quickly turn brown (**116, 117**). A chlorotic halo may then develop around the lesion. In more mature lesions, symptoms of the disease caused by *S. tritici* usually include the presence of black pycnidia (spore cases), which are visible to the naked eye (**118, 119**). Early symptoms of the disease on leaves are similar.

During prolonged humid weather, cirri of *S. tritici* are produced, which tend to be creamy-white (**120, 121**).

Disease cycle

All pathogens responsible for the disease survive inter-crop periods as dormant mycelium, pycnidia, and pseudothecia on seed, stubble, debris, and over-wintering cereal crops. Initial infections arise from wind-borne ascospores released from pseudothecia and asexually produced water splash-dispersed pycnospores produced from pycnidia. Usually the latter are responsible for disease epidemics. *S. tritici* is favoured

114. Leaf blotch (*Septoria tritici*) lesions on wheat flagleaf.

115. Leaf blotch (*Septoria tritici*) lesions on wheat leaf.

116. Leaf blotch (*Septoria tritici*) infection on a young wheat plant.

117. Extensive leaf blotch (Septoria tritici) lesions on leaves of maturing wheat.

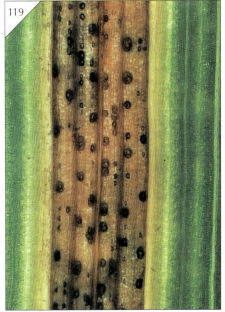

118. Photomicrograph of leaf blotch (*Septoria tritici*) lesions showing the pycnidia.

119. Photomicrograph of a single leaf blotch (*Septoria tritici*) lesion showing the pycnidia in more detail.

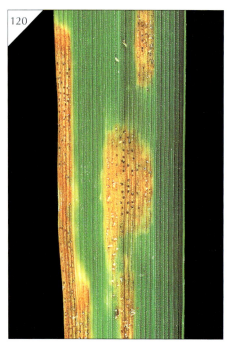

120. Leaf blotch (*Septoria tritici*) lesions showing pycnidia and cirri.

121. Photomicrograph of leaf blotch (*Septoria tritici*) lesion showing both pycnidia and cirri.

by temperatures of between 15 and 20°C. The shortest latent period for *S. tritici* it is between 21 and 28 days.

Economic importance

It is generally understood that Septoria leaf and glume blotch are among the most serious diseases of cereals, particularly in maritime cereal-growing areas. In recent years in the United Kingdom *S. tritici* has predominated and national average yield losses of £18 million have been estimated as a result of the disease. This is based on crop yields after fungicides have been applied. Severe glume blotch can reduce both 1000-kernel weight and specific weight (test weight), but may result in an improvement in protein content of grain. In the United States, average yield losses caused by Septoria leaf blotch were 0.41% for each 1% increase in leaf blotch on three wheat cultivars.

Control

Cultural control of these diseases includes disposal of contaminated crop debris by burning or ploughing. Crop rotation in which cereals occur every third year may reduce carry-over of inoculum.

Genetic resistance in winter wheat varieties is important in control of the disease, but only a moderate degree of resistance is exhibited in the field and several popular varieties in Europe are particularly susceptible to *S. tritici*.

Chemical control of the disease is widely practised in Europe. Application of azole-based fungicides, usually to the flag leaf of wheat, can be highly effective in reducing the disease. However, more than a single application of fungicide may be necessary during very wet seasons. There is widespread resistance to the benzimidazole (MBC) group of fungicides in United Kingdom populations of *S. tritici*.

Stagonospora nodorum (ex. *Septoria nodorum*) (sexual stage: *Phaeosphaeria nodorum*)

Diseases: Septoria Leaf and Glume Blotch

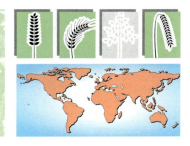

Septoria leaf and glume blotch is a disease primarily affecting wheat in most cereal-growing areas of the world, particularly during wet summers. Barley may also be affected by *Stagonospora nodorum* and Septoria leaf blotch on barley has been attributed to *Septoria passerinii*.

Symptoms of the disease caused by *S. nodorum* are usually distinguished from those of *S tritici* by the absence of black pycnidia. *S. nodorum* readily colonizes wheat ears in wet summers and the purple-brown symptoms of glume blotch then appear (**122, 123**). On mature leaves, *S. nodorum* tends to produce oval lesions that coalesce to form large areas of dead brown tissue (**124**). The pycnidia produced by *S. nodorum* are brown and not easy to see in a brown lesion (**125**).

During prolonged humid weather the pycnidia of *S. nodorum* exude salmon-pink cirri (spore masses).

122. Leaf blotch (*Stagonospora nodorum*) on a maturing wheat crop.

123. Glume blotch (*Stagonospora nodorum*) on a wheat ear.

124. Leaf blotch (*Stagonospora nodorum*) lesions on wheat leaves.

125. Photomicrograph of a leaf blotch (*Stagonospora nodorum*) lesion.

Disease cycle

All pathogens responsible for the disease survive inter-crop periods as dormant mycelium, pycnidia and pseudothecia on seed, stubble, debris, and over-wintering cereal crops. *S. nodorum* may also survive on wild grasses. Initial infections arise from wind-borne ascospores released from pseudothecia and asexually produced water splash-dispersed pycnospores produced from pycnidia. Usually the latter are responsible for disease epidemics. Temperatures of 20–27°C, together with prolonged high humidity (6–16 hours), are optimal for spore production and germination in *S. nodorum*. The shortest latent period for *S. nodorum* is between 10 and 14 days. Spores of *S. nodorum* may be splashed to ears, particularly during heavy rainfall, resulting in glume blotch.

Economic importance

It is generally understood that Septoria leaf and glume blotch are among the most serious diseases of cereals, particularly in maritime cereal-growing areas (see *S. tritici*).

Control

Cultural control of these diseases includes disposal of contaminated crop debris by burning or ploughing. Crop rotation in which cereals occur every third year may reduce carry-over of inoculum.

Genetic resistance in winter wheat varieties is important in control of the disease, but only a moderate degree of resistance is exhibited in the field.

Chemical control of the disease is widely practised in Europe. Initially, seed infection, usually by *S. nodorum*, can be reduced by application of fungicide seed treatments containing azole components. Application of azole-based fungicides, usually to the flag leaf of wheat, can be highly effective in reducing the disease. However, more than a single application of fungicide may be necessary during very wet seasons.

Snow moulds

Microdochium nivale (syn. *Fusarium nivale*), asexual stage; *Monographella nivalis*, sexual stage

Disease: Pink Snow Mould

Pink snow mould is a disease of winter wheat, barley, oats, rye, and many cultivated grasses that occurs in areas where snow falls on unfrozen soil and persists through much of the winter. The disease is widespread in wheat-growing areas of the northern hemisphere including Canada, parts of the former USSR, Japan, Scandinavia, Central and Eastern Europe, Scotland, England, and parts of the northern United States, including Alaska.

Symptoms of pink snow mould are apparent only after snow melt (**126, 127**). Initially, infected plants have a whitish covering of mycelium and sporodochia of the pathogen, which soon turns a characteristic salmon pink (**128, 129**). Infected leaves and leaf sheaths remain intact (as opposed to disintegrating),

become dry, and have a light- to dark-brown colour. Disease severity ranges from relatively small, discrete lesions on leaves to complete destruction of the foliage and dead plants. When snow is not present, the pathogen is restricted to leaf sheaths in contact with the soil and disease appears as superficial necrotic lesions. Dark-coloured fruiting structures (perithecia) may form later in the spring on plants with superficial infections.

Disease cycle

Microdochium nivale survives as hyphae and/or perithecia in infested host residue between susceptible crops. Infection of leaf sheaths and blades near the soil surface results from hyphae growing from perithecia or

126. Symptoms of pink snow mould in a field of wheat.

127. Symptoms of pink snow mould in barley.

128. Sporodochia and mycelia on barley leaves with pink snow mould.

infected residue. Infection and initial colonization occur during cool, wet weather and continue under snow. Dead plants and plant material returned to the soil after snow melt complete the disease cycle.

The role of ascospores as inoculum to initiate pink snow mould is questionable since most are discharged in the spring or summer. However, ascospores may be an important source of inoculum for head scab. *Microdochium nivale* can be seed borne and cause seedling blight. However, the role of seed-borne inoculum for pink snow mould is unknown.

Pink snow mould is most severe during years with wet and cool autumn weather followed by persistent snow on unfrozen soil. Virulence of *M. nivale* decreases at temperatures below 5°C and, for this reason, infection and initial colonization are believed to occur in the autumn before snowfall.

Economic importance

Accurate estimates of yield loss are not available for pink snow mould. Crop damage varies among years, ranging from complete destruction of the foliage and dead plants to spotty patches within a field with only minor loss.

Control

Cultural practices such as crop rotation provide time for infected crop debris to decompose and thus to reduce inoculum. Controlling weeds during the rotation is important since many grasses are hosts for *M. nivale*. Sowing winter cereals relatively early in the autumn results in larger plants better able to tolerate the disease.

In general, rye is most resistant to pink snow mould, followed by wheat and barley. Disease-resistant cultivars are available for rye and wheat, but not for winter barley. Resistance in wheat to pink snow mould is cor-

129. Foliar and root symptoms of pink snow mould on barley seedlings..

related with resistance to speckled snow mould caused by *Typhula idahoensis*.

Benzimidazole fungicides applied to the foliage before snowfall can reduce pink snow mould. Seed treatments with benzimidazole or azole fungicides control seed-borne inoculum (see Fusarium seedling blight, foot rot, and head blight).

Myriosclerotinia borealis (syn. *Sclerotinia borealis, S. graminearum*)

Disease: Snow Scald

Snow scald is limited to areas where winter cereals are grown in the far northern latitudes including Japan, Russia, northern Europe (Sweden, Finland, and Norway), and the United States (Alaska, Washington, and Minnesota). Snow scald may occur as part of a snow mould complex including *Microdochium nivale* and *Typhula* sp. The disease occurs on wheat, barley, rye, triticale, and several wild and cultivated grasses.

Symptoms of snow scald are first visible as irregular patches of dead plants in the field after snow melts in the spring. Infected plants are covered with white-to-grey mycelium that become grey to brown when dry. Leaves on infected plants become twisted and stringy (**130**) as the infected tissue breaks apart, and are covered with numerous black, irregularly shaped sclerotia approximately 2–15 mm long (**131**).

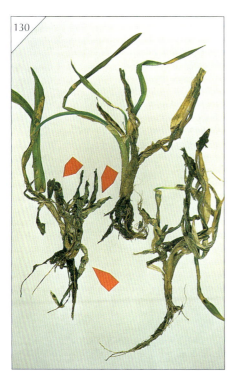

130. Symptoms of snow scald and sclerotia of the pathogen on wheat seedlings (courtesy of Professor J. Drew Smith, University of Saskatchewan).

131. Germinated sclerotia with apothecia of *Myriosclerotinia borealis* (courtesy of Professor J. Drew Smith, University of Saskatchewan).

Infected plants that are not already killed by the fungus may continue to die after snow melt (**132**).

Disease cycle

Sclerotia of *Myriosclerotinia borealis* persist between crops in soil and germinate in the autumn when wet weather and cool (below 10°C) temperatures occur, forming small (4–9 mm diameter), light-brown apothecia. Ascospores are ejected and wind disseminated to the leaves of cereal plants, which they infect. Hyphae colonize leaves and crowns under the snow. Sclerotia are formed from the hyphae during colonization of the plant.

Snow scald is favoured by wet and cold autumn weather, slightly frozen soil, and deep snow that persists for more than 100 days. The extent of host colonization is determined by the length of snow cover.

Economic importance

Accurate estimates of yield loss associated with snow scald are lacking. However, reports exist of individual fields with up to 70% of plants dead because of snow scald.

132. Symptoms of snow scald on rye (courtesy of Professor J. Drew Smith, University of Saskatchewan).

Control

Cultural practices including adequate fertility, especially phosphorus, may reduce the damage associated with snow scald. Practices such as spreading coal dust to encourage snow melt can also reduce disease development.

Disease resistance is the only practical control measure. Winter wheat is more susceptible to snow scald than winter rye; however, cultivars of both vary in their reaction to the pathogen. Resistance to snow scald in wheat is correlated with cold hardiness.

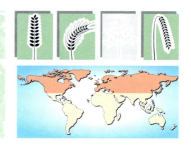

Typhula idahoensis, *T. ishikariensis*, and *T. incarnata* (syn. *T. itoana*)

Disease: Speckled Snow Mould

Speckled snow mould occurs only in areas where winter cereals are grown with snow cover that persists for at least 100 days. This disease has been reported in Canada, some northern states of the United States. (including Alaska), Japan, Northern and Eastern Europe, and parts of the former USSR. In addition to speckled snow mould, *Typhula incarnata* is capable of causing a root and crown rot in the absence of snow cover (**133**) and, consequently, is more widely distributed than *T. idahoensis* and *T. ishikariensis*. These pathogens cause disease on all winter cereals and many grasses. *T. idahoensis* and *T. ishikariensis* are also able to infect some legume hosts, such as *canola* (oilseed rape) and clover, grown in rotation with winter cereals.

Symptoms and signs of speckled snow mould are apparent following snow melt (**134**). Foliage of infected plants is matted to the soil and covered with a whitish-grey mycelium. The mycelium disappears with a few days of dry weather. Numerous dark-coloured sclerotia are present over the surface of infected plants (**135, 136**). Sclerotia of *T. idahoensis* and *T. ishikariensis* are spherical (0.3–2 mm) and dark-brown to black. In contrast, sclerotia of *T. incarnata* are irregularly shaped (0.5–5 mm), reddish-brown, and are more abundant on roots and between sheaths in the crown (**137, 138**) than *T. idahoensis* and *T. ishikariensis*. Disease severity ranges from patches to complete destruction of all above-ground foliage within a field (**139**).

133. Symptoms of snow mould caused by *Typhula incarnata* on barley.

134. Symptoms of snow mould on wheat leaves, as snow recedes (courtesy of Professor Robert L. Forster, University of Idaho).

135. Sclerotia of *Typhula* spp. on wheat leaves with snow mould (courtesy of Professor Robert L. Forster, University of Idaho).

136. Sclerotia of *T. incarnata* on barley leaves.

Disease cycle

These fungi survive between crops as sclerotia in soil and infected host debris. Sclerotia germinate in the autumn during cool and wet weather, producing short (5–25 mm) sporophores from which basidiospores are liberated. Sporophores of *T. idahoensis* are tan to brown; those of *T. ishikariensis* are white to tan with lavender tints, and those of *T. incarnata* are light pink. Most infections are the result of hyphae growing from sporophores or directly from sclerotia in the soil under snow. It is unclear exactly when germination of sclerotia and infection begins under snow, but colonization of infected plant tissues continues as long as snow cover persists.

Development of speckled snow mould depends on relatively deep snow cover that persists for at least 100 days. Disease is more severe in years when abundant rain falls during

137. Sclerotia of *T. incarnata* on the stem base of a young barley plant.

138. Sclerotia of *T. incarnata* on barley crown.

139. Total destruction of wheat crop by speckled snow mould (courtesy of Professor Robert L. Forster, University of Idaho).

autumn and when snow falls on unfrozen or slightly frozen soil.

Economic importance

The occurrence of speckled snow mould is sporadic. In years when disease is severe, entire fields may be killed and require resowing, whereas, in other years, the disease does not occur or is insignificant.

Control

Disease-resistant cultivars are available for wheat in some areas. Rye is inherently more resistant than wheat and is grown in some areas where resistant cultivars are not available or disease is especially severe.

Cultural practices such as sowing late in the summer or early in the autumn promote the development of large, well-tillered plants better able to tolerate the disease. Spreading dark-coloured materials such as coal dust on fields hastens snow melt and has been used to limit damage due to speckled snow mould.

Azole fungicides applied to the foliage in the late autumn before snow cover can control disease. Seed treatment with azole fungicides may provide some control of speckled snow mould, but is not practised.

Stripes

Cephalosporium gramineum
(syn. *Hymenula cerealis*)

Disease: Cephalosporium Stripe

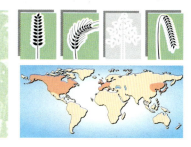

Cephalosporium stripe is a vascular disease of winter cereals grown in temperate regions, especially areas where soil freezes during the winter. The disease occurs in Japan, the United States, where it is especially important in the Pacific Northwest (Idaho, Oregon, Montana, and Washington) and Great Plains (Kansas), the United Kingdom (England and Scotland), and other parts of Europe. All of the winter cereals are hosts for this pathogen, as are several grass species; however, wheat is the most important host.

Symptoms of Cephalosporium stripe usually appear in the early spring after plants have resumed growth. The most prominent symptoms include long yellow stripes in the leaf blades that extend down the leaf sheath (**140**). Close examination of the yellow stripes often reveals the presence of brown streaks in individual vascular bundles (**141, 142**). For Cephalosporium stripe, one diagnostic feature is the presence of symptoms in each leaf blade and sheath below the uppermost leaf with symptoms. Plants are stunted (50–75% the height of healthy plants) and a 'double canopy' with two prominent ear-bearing layers is apparent when disease is severe.

140. Chlorotic stripes typical of Cephalosporium stripe of wheat.

141. Chlorotic stripes in leaf blades and sheaths of wheat with Cephalosporium stripe.

142. Chlorotic stripes and necrotic streaks in wheat with Cephalosporium stripe.

Disease cycle

Cephalosporium gramineum is a facultative parasite that survives between susceptible crops as a saprophyte in residue from a previously infected crop; this fungus does not colonize residue incorporated into soil from subsequent crops. The fungus sporulates profusely during the autumn on infected residue near the soil surface and the single-celled conidia are washed into the soil. The precise events involved with infection are unclear; however, the fungus is capable of colonizing the roots, subcrown internodes, and possibly the stem bases where adventitious roots emerge. Root injury incurred during the winter is believed to allow the pathogen to enter the xylem, where it sporulates and spreads throughout the plant in the transpiration stream. *C. gramineum* remains in the xylem as long as the plant is alive; however, the fungus colonizes plant tissues surrounding the xylem as the plant senesces and is returned to the soil in the now-colonized host debris following harvest.

Cool (5–10°C) temperatures and rainfall during the autumn are favourable to sporulation of *C. gramineum* and result in abundant inoculum. Low temperatures during winter that promote soil freezing, especially alternating soil freezing and thawing, are favourable to disease development. Cultural practices such as early seeding of winter cereals in the autumn and high fertility levels result in large plants that are more susceptible to winter root injury and, thus, favour Cephalosporium stripe.

Economic importance

Cephalosporium stripe has the potential for extreme destruction: loss in grain yield for stems with stripes extending into the ear is estimated at 85% for very susceptible cultivars. Grain from diseased plants is also shrivelled and test weight (specific weight) is reduced significantly, rendering it unsaleable or suitable only for animal feed.

Control

Cultural control of Cephalosporium stripe involves delayed sowing of winter cereals in the autumn, which results in smaller plants with fewer roots that are susceptible to winter root injury. Increasing the length of time between winter cereals provides time for infected crop residue to decompose, thus eliminating the pathogen. The length of rotation depends upon the climate: longer rotations are required in areas with arid summers owing to reduced rates of straw decomposition. In contrast, infected straw decomposes faster in areas with moist summers and thus, rotations may be shorter. However, neither of these practices is completely effective in controlling the disease.

Genetic resistance offers the most promising method of control for Cephalosporium stripe. Currently, cultivars with some tolerance of the pathogen are available, but cultivars with highly effective resistance do not exist in wheat or barley. Wheat germ plasm with highly effective resistance to Cephalosporium has not been identified. Efforts are under way to transfer resistance from wheat grass (*Agropyron* sp.) to wheat.

Chemical control of Cephalosporium stripe has not been effective.

Pyrenophora graminea (asexual stage: *Drechslera graminea*)

Disease: Barley Leaf Stripe

This seed-borne disease occurs wherever barley is grown, but is rare where effective seed treatments are used. An almost worldwide ban on the use of organomercurial seed treatments has resulted in an increased interest in the problem. Affected seedlings are usually stunted and may suffer from pre- and post-emergence death. Usually, symptoms appear on the second and third leaves to develop. One or more long chlorotic stripes appear parallel to the leaf rib and often extend the whole length of the leaf (**143, 144**). Affected areas may become necrotic and tear, resulting in a shredded appearance. As plants mature, ears may not emerge from the sheath, or they may emerge as blighted, twisted, compressed and brown (**145**). Grain production in affected plants is severely reduced.

Disease cycle

Pyrenophora graminea is exclusively seed borne and can go through only one cycle of infection during a season. The fungus exists as mycelium in the seed coat and pericarp, but not in the embryo. Infection of the seedling is influenced by temperature and humidity. Soil temperatures below 10°C during germination favour infection, whereas temperatures above 12°C reduce it. High humidity during the period of anthesis results in sporulation on affected leaves.

143. Barley leaf stripe symptoms on barley crop.

144. Barley leaf stripe on the leaves of a maturing barley crop.

145. Leaf symptoms and aborted ear from infected barley plant.

Conidia are wind blown to heads where infection of seed occurs most frequently during early grain development. Free water is not necessary for infection. The sexual ascospores produced in pseudothecia are rare in the field and are not considered important in the disease cycle.

Economic importance

It has been estimated that, since affected plants produce very little grain, yield decreases in the range 0.5–1% for each 1% of plants affected by leaf-stripe. Prior to the advent of highly effective seed treatments, yield losses in barley crops were on occasion very high.

Control

Cultural control practices include the use of good-quality pathogen-free seed. Resistant varieties are available, but there is evidence of geographically specific races of the pathogen. Consequently, varieties may differ in their resistance, according to where they are grown.

Chemical control was relied upon for many years by using seed treatments containing organomercurial compounds. However, there were reports of resistance in populations of *P. graminea* to organomercury in the United Kingdom during the mid 1980s. Since the early 1990s, the use of organomercury has almost ceased in Europe, for environmental and toxicity reasons. Control of leaf stripe is currently achieved effectively by inclusion of the azole imazalil or triazoxide in seed treatments.

Stem Base and Root Diseases

Fungi

Cochliobolus sativus, sexual stage; *Bipolaris sorokiniana* (syn. *Helminthosporium sativum*, *H. sorokinianum*), asexual stage

Disease: Common root rot; Seedling blight (Spot blotch)

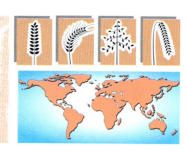

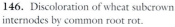

146. Discoloration of wheat subcrown internodes by common root rot.

147. Subcrown internode of wheat with common root rot.

As indicated by the common name of this disease, all plant parts are susceptible to infection by *C. sativus* depending upon the prevailing weather conditions. Symptoms of seedling blight include brown, elliptical lesions that progress inward and upward from near the base of the coleoptile below the soil surface. Plants may die before emergence, but usually die after emergence. Common root rot is characterized by dark-brown to black, necrotic lesions on roots, subcrown internodes, and stem bases (**146, 147**). Discoloration of the subcrown internode is characteristic of infection by *C. sativus*. Lesions often coalesce forming large areas of necrotic tissue in the crown.

For further details see Leaf and Stem Diseases, Spots/Blotches.

***Fusarium avenaceum* (*Gibberella avenacea*),
F. culmorum, *F. graminearum* (*Gibberella
zeae*), *F. poae*, *Microdochium nivale* (formerly
F. nivale) (*Monographella nivalis*)**

Disease: Fusarium seedling blight; Foot rot;
Dryland foot rot

148. Foot rot damage and mycelium on wheat
stem base.

149. Foot rot damage to wheat stem base.

Fusarium foot rot (crown rot, brown foot rot)
symptoms are varied. In temperate areas, the
most common symptom is a dark brown
lesion around the node of mature plants.
Long thin dark brown vertical streaks are also
frequently observed. In more arid areas, dryland foot rot may develop. The entire stem
base becomes girdled with a dark brown
lesion (**148, 149**). Tissue may become soft
and white, or pink fungal growth with orange
spore masses can develop.

Disease cycle

All of the species implicated in the disease can
survive saprophytically in the soil or on plant
material of a range of different crops and weed
species. In addition, they can all be seed-borne
on cereals. It is possible that seedlings which
survive the initial infection may develop foot
rot at a later stage of growth. Environmental
conditions are likely to influence disease development, with moisture stress resulting in
severe symptoms of dryland foot rot caused by
F. culmorum and *F. graminearum*. Under
such conditions, sporulation may occur on
stem bases and nodes.

Economic importance

The economic importance of Fusarium foot
rot is difficult to determine for three reasons.
First, the effect of each individual pathogen on
yield may vary. Second, naturally occurring

foot rot may be the result of multiple infection by two or more *Fusarium* species alongside other important stem base pathogens, such as *Pseudocercosporella herpotrichoides* (eyespot). Finally, as yet there are no highly effective and reliable fungicides to treat the disease. Recent unpublished work has shown that foot rot caused by *F. culmorum* and *M. nivale* may reduce yield by over 30%.

Control

Cultural control of the Fusarium diseases includes disposal of contaminated debris and crop rotation. Fusarium ear blight has been shown to be severe following maize.

Chemical control of Fusarium foot rot is inconsistent. Benzimidazoles and azoles are sometimes used, but most isolates of *M. nivale* in the UK are resistant to former fungicides.

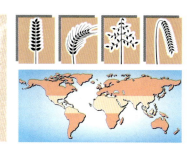

Gaeumannomyces graminis var. ***avenae*** (oats)
Gaeumannomyces graminis var. ***graminis***
(wheat, barley, rye)
Gaeumannomyces graminis var. ***tritici*** (wheat,
barley, rye)

Disease: Take-All

Take-all is mainly a problem in autumn-sown wheat or barley in temperate cereal-growing areas. Oats can be affected by a specific pathotype of *Gaeumannomyces graminis*, but it is uncommon. Rhizomatous grasses can also harbour the pathogen. The first symptoms of severe take-all may occur in young plants, which become yellow and stunted. Characteristic symptoms are seen on more mature plants, particularly after a dry period. Patches of stunted, prematurely ripe plants become apparent (**150**), bearing whiteheads (small, bleached ears with little or no grain) (**151**). Roots of affected plants are blackened and stunted (**152**). The stem base may darken (**153**) and, especially during wet weather, perithecia may be seen as small dark spots.

Disease cycle

The main source of inoculum for take-all is mycelium on contaminated roots or cereal debris. Infected grasses may also provide a source of inoculum. Although the pathogen can, under certain circumstances, reproduce sexually, resulting in rain- and wind-dispersed ascospores, they are considered unimportant under field conditions. The main infection of cereal roots occurs when temperatures exceed 10°C. Mycelia grow from a food source onto cereal roots and then spread along roots by producing long runner-hyphae. Such hyphae can also result in plant-to-plant spread of disease. Periodically along the runner-hyphae, loose gatherings of hyphae called hyphopodia occur and infection pegs develop beneath these, through which the fungus feeds. As the season progresses, more of the roots are affected and the stem base may also be colonized. Severe root rot deprives the plant of water and nutrients, resulting in premature ripening and whiteheads. After harvest the pathogen remains in the roots and on cereal

150. Take-all patches in wheat crop in ear.

151. White heads caused by take-all infection of wheat plants.

152. Blackened roots and stem bases on take-all infected wheat plants.

153. Adventitious roots developing to compensate for take-all infection on wheat plants.

debris, and will remain viable until the residues decay.

Take-all is most severe in light, loose, alkaline, infertile soils. Poor drainage, early sown crops, and continuous cereals also exacerbate the problem.

Economic importance

In the United Kingdom wheat crop, take-all is considered to be the most important disease pathogen in terms of economic loss, with yield reduction estimated at £37 million per year. Yield losses are generally greatest in second and third successive wheat crops, where 10–20% of the yield may be lost. Yield losses of 5–10% may go unnoticed if symptoms are not obvious. Severe take-all can significantly reduce 1000-grain weight and specific weight (test weight) of grain.

Control

Cultural control currently offers the only satisfactory means of reducing take-all. A 1-year break from susceptible cereal crops together with effective rhizomatous grass weed and volunteer cereal control, will significantly reduce the problem. Other cultural practices can also influence the disease. Seed should be drilled at an optimum time into firm, well-drained seedbeds and a balanced fertilizer programme should be adopted. After several (two to five) consecutive cereal crops, the severity of take-all usually diminishes. This is called take-all decline. This phenomenon is not well understood, although it is generally considered to be a natural form of biological control. Microorganisms antagonistic to *G. graminis* may build up in the soil.

Disease resistance to take-all is low in wheat and barley varieties. It is possible, however, to grow oats in most areas as a break crop because of the rarity of *G. graminis* var. *avenae*. There is also some evidence for resistance to take-all in some varieties of triticale.

Chemical control of the disease has been unsuccessful to date. Some azole-based systemic seed treatments may suppress the disease in its early stages, and products in development offer some promise.

Pseudocercosporella herpotrichoides (sexual stage: *Tapesia yallundae*)

Disease: Eyespot (Foot Rot, Strawbreaker)

Eyespot, sometimes referred to as foot rot or strawbreaker, is a disease primarily of autumn-sown wheat and barley in cooler maritime wheat-growing areas of the world, including northern United States, parts of South America, Europe, New Zealand, Africa and Australasia. Early symptoms of the disease on young plants often consist of indistinct honey-brown lesions on the stem base (**154**). Occasionally, one or two leaf sheaths may be penetrated by the pathogen, resulting in a small black pinprick. Characteristic symptoms of the disease occur on stem bases of more mature plants. Eye-shaped honey-brown lesions develop, generally below the first node. Such lesions have a diffuse margin (cf. sharp eyespot) and a central black 'pupil' consisting of a mass of compacted hyphae that is difficult to remove by rubbing (**155**). Grey mycelium may develop in the stem cavity and severe penetrating lesions can result in plants breaking at the lesion and falling over (lodging) (**156**). Severe eyespot can also cause the production of bleached, prematurely ripe ears containing little or no grain (whiteheads).

Disease cycle

There are two sources of inoculum for eyespot disease. Probably the most common is long, thin asexual spores, produced mainly on cereal stubble and debris during autumn, through mild winters, and into early spring. However,

154. Eyespot lesion on maturing wheat plant.

155. Eyespot lesion on the stem base of a young barley plant.

156. Wheat crop lodged after severe eyespot infection.

the sexual stage of the pathogen, *Tapesia yallundae*, has been identified recently in many cereal-growing areas including Australia, Germany, and the United Kingdom, and it has been proposed that ascospores may be produced over much of the growing season if conditions are conducive. The significance of ascospore inoculum in the epidemiology of the disease has yet to be fully evaluated. Conidia are rain-splash dispersed over short distances to stem bases of cereals where prolonged humid, cool (5–16°C) weather is conducive to infection. Following infection, symptoms may not be visible for several weeks or months, depending on environmental conditions. The disease is predominantly monocyclic, although under optimal conditions for infection and sporulation, secondary inoculum may be produced. Moisture-retentive, heavy clay soils are conducive to eyespot and early sown, over-fertilized crops are particularly prone to the disease.

Economic importance

It is generally understood that superficial disease does not affect yield significantly. However, when the pathogen has deeply penetrated the stem base, causing tissue degradation, considerable yield losses may occur. It has been estimated that the mean annual value of crop losses attributable to eyespot from 1985 to 1989 in the United Kingdom was over £26 million. In an evaluation of yield loss relationships for the disease in the United Kingdom, it was proposed that each 1% increase in the percentage of tillers affected by severe eyespot was associated with a yield loss of 0.21%. Lodging caused by eyespot slows down harvesting and can result in poor grain quality of high moisture content.

Control

Cultural control of eyespot includes disposal of contaminated stubble and debris by ploughing or burning (where permitted). A 2-year break from cereals may also reduce disease, as may a balanced approach to fertilizer use and sowing time. Lodging may be reduced by the application of a growth regulator.

Disease resistance in winter wheat varieties grown in Europe has been derived mainly from the variety Cappelle Desprez. More recently, improved resistance from *Aegilops ventricosa* has been introduced into some varieties.

Until the early 1980s, the benzimidazoles (MBCs) were used extensively in Europe for chemical control of the disease. The pathogen population developed widespread resistance to this group and now azoles are widely used in their place. Fungicides are usually applied at between growth stage (GS) 30 and GS 32. In the Pacific North-West region of the United States, benzimidazoles are still used, even though fungicide-resistant strains of the pathogen are widespread.

Rhizoctonia cerealis (sexual stage: *Ceratobasidium cereale*)

Disease: Sharp Eyespot

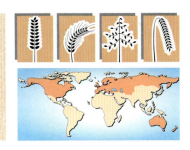

157. Sharp eyespot lesions on wheat stem bases.

Sharp eyespot is a disease primarily of autumn-sown wheat and barley, although oats and rye may also be affected. Sharp eyespot occurs in most temperate cereal-growing areas of the world, including Europe and North America. Occasionally, *Rhizoctonia cerealis* together with *R. solani* cause pre-emergence damping-off in cereals, especially if fields are waterlogged or particularly cold at sowing. In young plants, symptoms of sharp eyespot consist of indistinct stem-base browning, which can be very difficult to distinguish from eyespot and Fusarium foot rot. Characteristic symptoms of sharp eyespot occur on mature plants, which suffer from pale cream oval lesions with a dark-brown margin on basal leaf sheaths (**157–159**). Lesions are frequently superficial, but occasionally penetrate the stem, which results in tissue damage, lodging and whiteheads.

158. Sharp eyespot at the base of a wheat crop.

159. Sharp eyespot lesions on wheat stems.

Disease cycle

The major source of inoculum for sharp eye-spot is mycelia on infected stubble. The fungus can produce soil-borne sclerotia, which also provide an inoculum source. Hyphal infection of cereal roots and stem bases occurs during cool (around 10°C), moist conditions. Thereafter, plants in dry, well-drained acid soils suffer more with the problem. Although sexually produced basidiospores may occur, their significance in the epidemiology of the disease is unclear. At the end of the season, infected debris and sclerotia provide inoculum for successive cereal crops.

There is some evidence for an interaction between sharp eyespot and eyespot (*Pseudocercosporella herpotrichoides*). It has been suggested that establishment of one of the pathotypes of eyespot (the W-type) inhibits development of sharp eyespot. Fungicidal control of eyespot can lead to a significant increase in sharp eyespot.

Economic importance

In surveys of UK wheat, sharp eyespot was common in every year, but penetrating lesions, which can result in yield loss, were relatively uncommon. Average yield losses in British crops are estimated at 0.4% per annum.

Control

There are currently no consistent and highly effective means of control of sharp eyespot. Cultural control probably offers the best way of reducing disease severity. Late sowing and disposal of debris, together with crop rotation, may reduce disease.

Disease resistance in cereal varieties is available, but not particularly effective. Cereal species can vary in their susceptibility: rye and oats are probably most susceptible and wheat and barley least.

Currently, no fungicides are highly effective against the disease.

Pathogen Structures

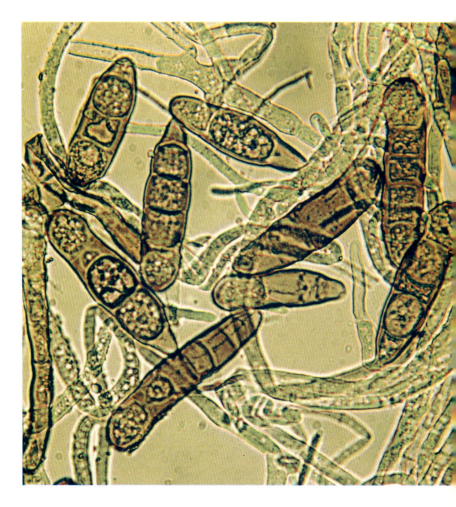

Introduction

Included in Section 4 are photographs of spores, fruiting bodies and hyphae of most of the fungal pathogen discussed in the previous Sections. These photographs are intended as an additional aid to diagnosis for those users who have access to basic microscopy facilities. Although a microscope is required to see most of the spores, some, such as the ergot stromata, are readily observable with the naked eye, and others, such as perithecia of *Gibberella zeae*, are visible with a good hand lens.

The organization of the photographs follows that of the main text. Types of disease are arranged within sections on the plant part affected – the Ear and Grain diseases, followed by the Leaf and Stem diseases and, finally, the Stem Base and Root diseases. Cross-references are given for pathogens that cause diseases on more than one part of the plant.

Ear and Grain Diseases

Blights

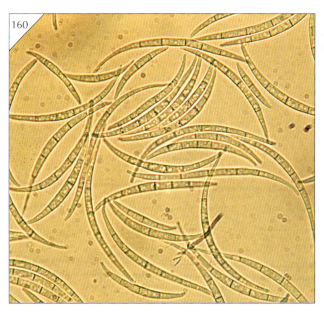

160. Macroconidia of *Fusarium avenaceum*, 400× (courtesy of CIMMYT).

161. Macroconidia of *Fusarium culmorum*, 400 × (courtesy of CIMMYT).

162. Macroconidia of *Fusarium graminearum*, 400× (courtesy of CIMMYT).

163. Perithecia of *Gibberella zeae* on wheat glume (courtesy of CIMMYT).

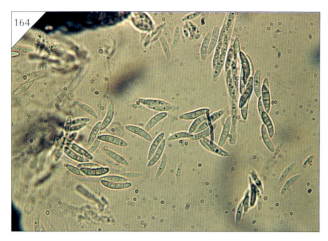

164. Asci and ascospores of *Gibberella zeae*, 400× (courtesy of CIMMYT).

165. Macroconidia of *Microdochium nivale*, 400× (courtesy of CIMMYT).

166. Perithecia of *Monographella nivalis* on a wheat stem base.

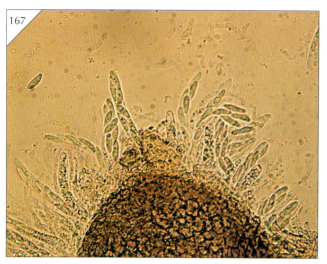

167. Asci and ascospores of *Monographella nivalis*, 400× (courtesy of CIMMYT).

Bunts/Smuts

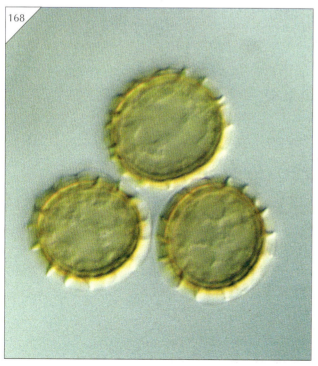

168. Teliospores of *Tilletia controversa* viewed with differential interference microscopy, 1000×.

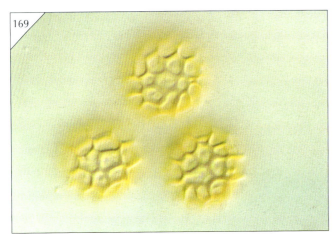

169. Teliospores of *Tilletia controversa* viewed with differential interference microscopy showing surface reticulations, 1000×.

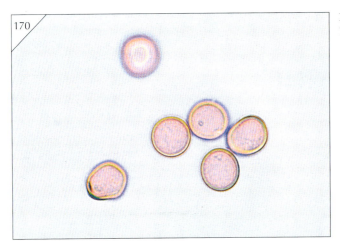

170. Teliospores of *Tilletia laevis*, 400×.

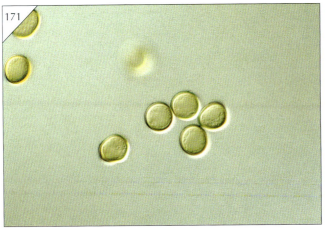

171. Teliospores of *Tilletia laevis* viewed with differential interference microscopy showing smooth spore surface, 400×.

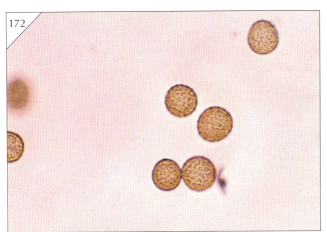

172. Teliospores of *Tilletia tritici*, 400×.

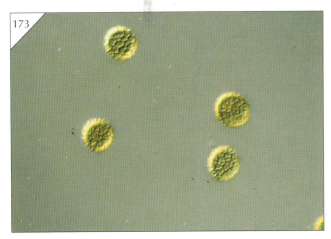

173. Teliospores of *Tilletia tritici* viewed with differential interference microscopy showing surface ornamentation, 400×.

174. Teliospores of *Tilletia indica*, 400×.

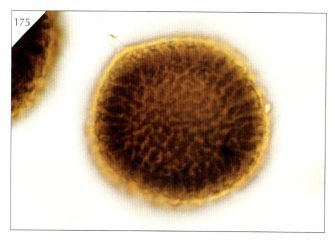

175. Teliospore of *Tilletia indica*, 1000×.

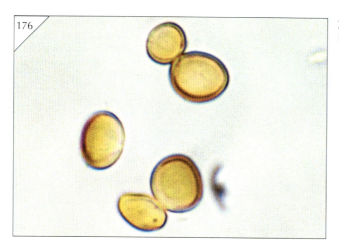

176. Teliospores of *Ustilago hordei*, 1000×.

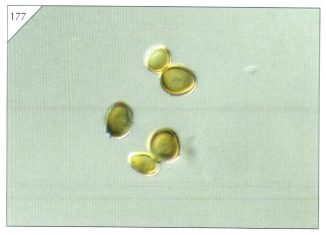

177. Teliospores of *Ustilago hordei* viewed with differential interference microscopy showing smooth spore surface, 1000×.

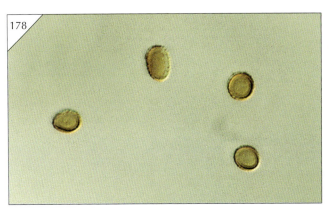

178. Teliospores of *Ustilago nuda* viewed with differential interference microscopy, 1000×.

179. Teliospores of *Ustilago nuda* viewed with differential interference microscopy showing surface ornamentation, 1000×.

Other Diseases

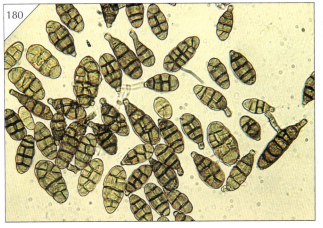

180. Conidia of *Alternaria* sp., 400× (courtesy of CIMMYT).

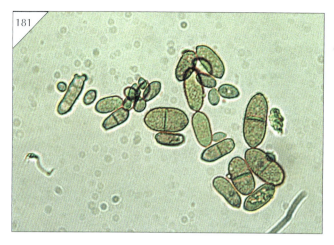

181. Conidia of *Cladosporium* sp., 400× (courtesy of CIMMYT).

182. Conidia of *Cladosporium* sp. viewed with differential interference microscopy, 1000×.

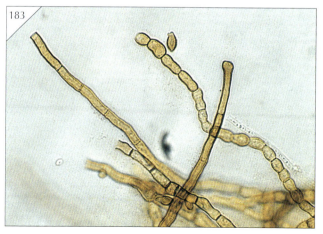

183. Hyphae and a conidium of *Cladosporium* sp., 400×.

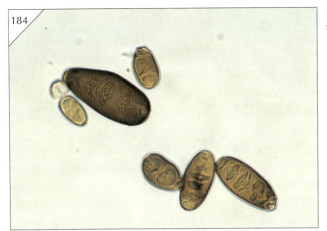

184. Conidia of *Bipolaris* sp., 400×.

185. Germinating sclerotium of *Claviceps* sp. with emerging stromata (courtesy of L.M. Carris).

186. Stromata of *Claviceps purpurea* (courtesy of CIMMYT).

187. Close-up of a stroma of *Claviceps* sp (courtesy of L.M. Carris).

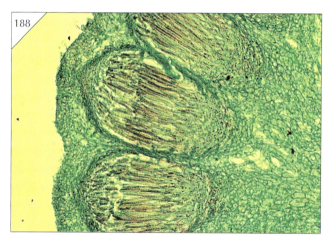

188. Cross-section through a stroma of *Claviceps purpurea* showing perithecia, 400×.

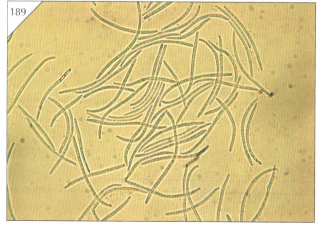

189. Conidia of *Septoria tritici*, 400× (courtesy of CIMMYT).

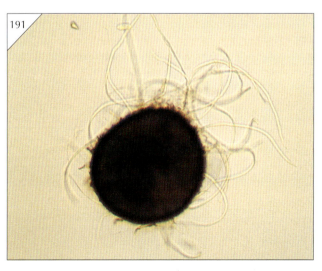

190. Conidia of
Stagonospora nodorum, 400×
(courtesy of CIMMYT).

Leaf and Stem Diseases

Mildew

191. Cleistothecium of
Blumeria graminis, 200×.

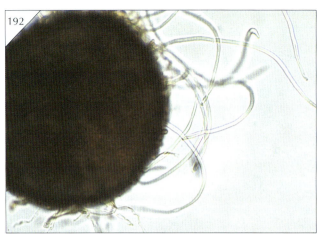

192. Cleistothecium of *Blumeria graminis* with hyphal appendages, 400×.

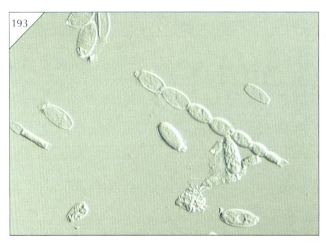

193. Conidia of *Blumeria graminis* viewed with differential interference microscopy, 200×.

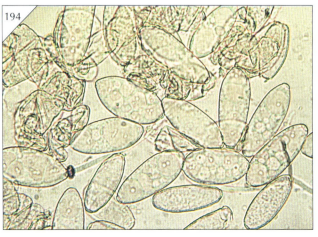

194. Conidia of *Blumeria graminis*, 400×.

Rusts

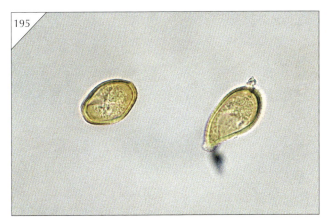

195. Urediniospores of *Puccinia coronata*, 400×.

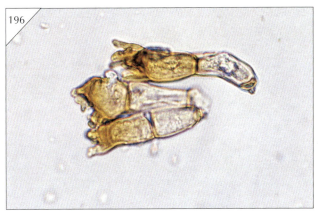

196. Teliospores of *Puccinia coronata* showing blunt projections from the teliospore, 400×.

197. Teliospores of *Puccinia coronata* viewed with differential interference microscopy showing blunt projections from the teliospore, 400×.

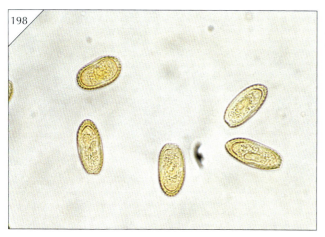

198. Urediniospores of *Puccinia graminis* f. sp. *tritici*, 400×.

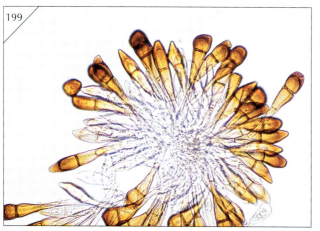

199. Teliospores of *Puccinia graminis* f. sp. *tritici*, 200×.

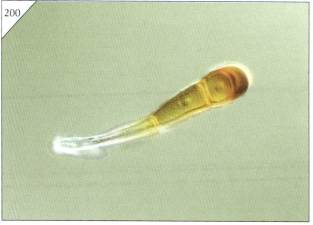

200. Teliospore of *Puccinia graminis* f. sp. *tritici* viewed with differential interference microscopy, 400×.

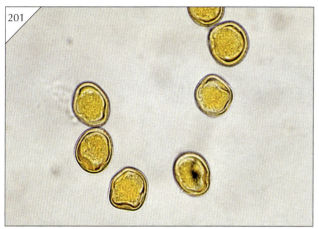

201. Urediniospores of
Puccinia recondita, 400×.

202. Teliospores of
Puccinia recondita, 200×.

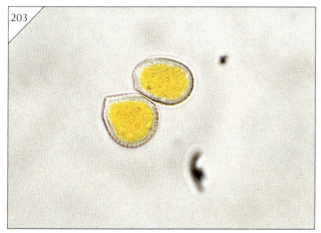

203. Urediniospores of
Puccinia striiformis, 400×.

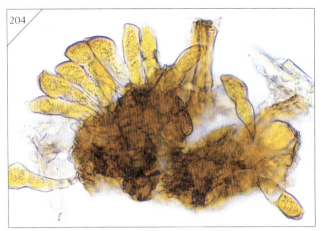

204. Teliospores of *Puccinia striiformis*, 200×.

Smuts

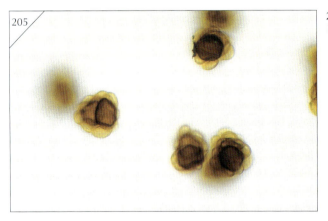

205. Teliospores of *Urocystis agropyri*, 400×.

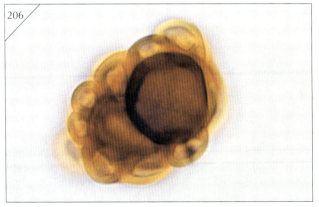

206. Teliospore of *Urocystis agropyri*, 1000×.

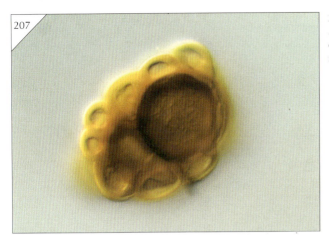

207. Teliospore of *Urocystis agropyri* viewed with differential interference microscopy, 1000×.

Spots/Blotches

208. Conidia of *Bipolaris sorokiniana*, 400× (courtesy of CIMMYT).

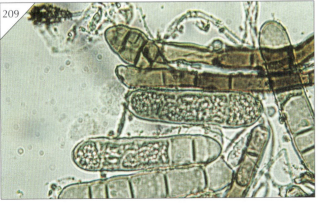

209. Conidia of *Drechslera teres*, 400× (courtesy of CIMMYT).

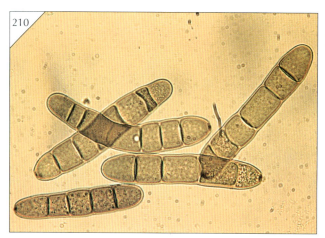

210. Conidia of *Drechslera avenae*, 400× (courtesy of CIMMYT).

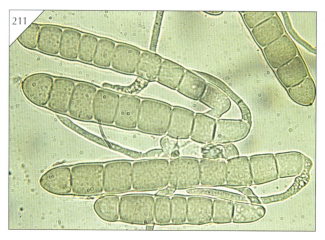

211. Conidia of *Drechslera tritici-repentis*, 400× (courtesy of CIMMYT).

212. Perithecia, asci, and ascospores of *Pyrenophora tritici-repentis*, 400× (courtesy of CIMMYT).

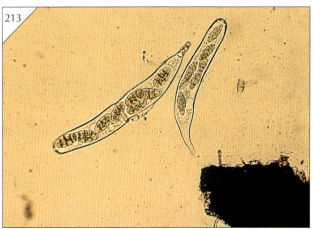

213. Asci and ascospores of *Pyrenophora tritici-repentis*, 400× (courtesy of CIMMYT).

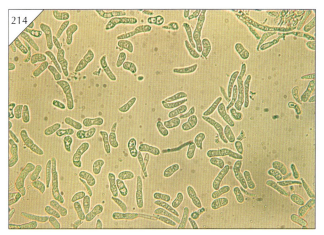

214. Conidia of *Rhynchosporium secalis*, 400× (courtesy of CIMMYT).

215. Conidia of *Selenophoma donacis* viewed with differential interference microscopy, 400×

216. Cirri of *Septoria tritici* emerging from pycnidia (courtesy of W.W. Bockus).

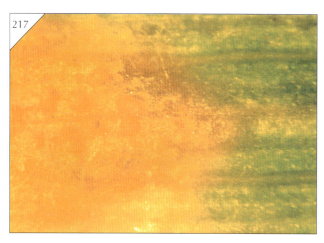

217. Cirri of *Stagonospora nodorum* emerging from pycnidia (courtesy of W.W. Bockus).

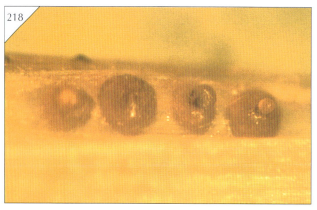

218. Conidia oozing from pycnidia of *Stagonospora nodorum* (courtesy of W.W. Bockus).

219. Ascospores of *Phaeosphaeria nodorum*, 400× (courtesy of CIMMYT).

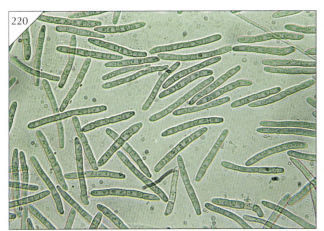

220. Conidia of *Stagonospora avenae*, 400× (courtesy of CIMMYT).

221. Asci and ascospores of *Phaeosphaeria avenaria*, 400× (courtesy of CIMMYT).

Snow Moulds

For *Microdochium nivale*, see Ear and Grain Diseases (blight).

Stripes

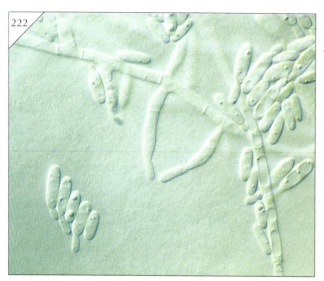

222. Conidia and conidiophores of *Cephalosporium gramineum* viewed with differential interference microscopy, 1000×.

223. Sporodochia of *Cephalosporium gramineum* on infested wheat straw (courtesy of W.W. Bockus).

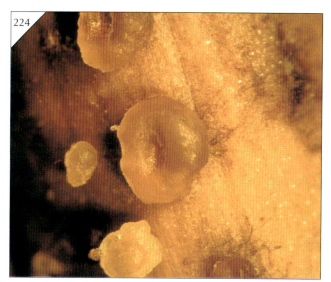

224. Close-up of sporodochia of *Cephalosporium gramineum* on infested wheat straw (courtesy of W.W. Bockus).

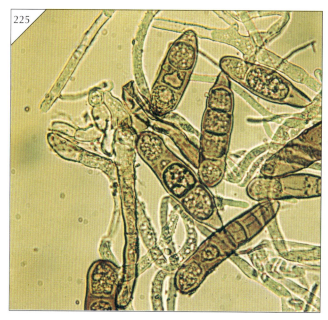

225. Conidia of *Drechslera graminea*, 400× (courtesy of CIMMYT).

Stem Base and Root Diseases

For *Bipolaris sorokiniana*, see Leaf and Stem Diseases (spots/blotches).

For *Fusarium avenaceum*, see Ear and Grain Diseases (blight).

For *Fusarium culmorum*, see Ear and Grain Diseases (blight).

For *Fusarium graminearum*, see Ear and Grain Diseases (blight).

For *Microdochium nivale*, see Ear and Grain Diseases (blight).

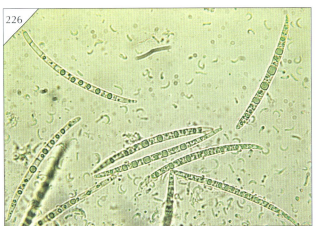

226. Ascospores of *Gaeumannomyces graminis* var. *tritici*, 400× (courtesy of CIMMYT).

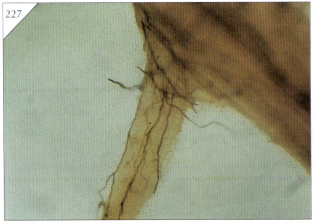

227. Runner hyphae of *Gaeumannomyces graminis* var. *tritici* on wheat root (courtesy of J.W. Sitton).

228. Runner hyphae of *Gaeumannomyces graminis* var. *tritici* on wheat root (courtesy of J.W. Sitton and D.M. Weller).

229. Conidia of *Pseudocercosporella herpotrichoides* viewed with differential interference microscopy, 400×.

230. Apothecia of *Tapesia yallundae* on straw piece (courtesy of P.S. Dyer).

231. Close-up view of apothecia of *Tapesia yallundae* (courtesy of P.S. Dyer).

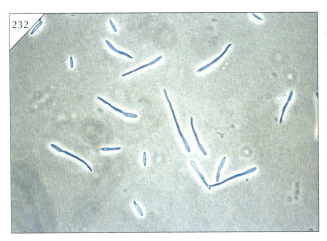

232. Germinating ascospores of *Tapesia yallundae* (courtesy of P.S. Dyer).

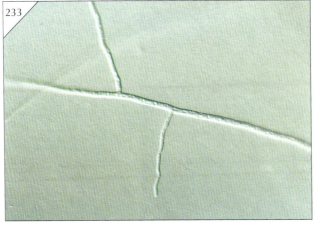

233. Hyphae of *Rhizoctonia solani* viewed with differential interference microscopy showing typical right angle branching, 400×.

Glossary of terms

Acervulus (pl. **acervuli**) A sub-epidermal, cushion-like mass of hyphae containing asexual conidia and conidiophores.

Active ingredient The active component of a formulated product.

Adult plant resistance Resistance detectable at the post-seedling stages of development (mature plant resistance). See also field resistance.

Aeciospore An asexually produced dikaryotic rust spore found in an aecium.

Aecium (pl. **aecia**) A cup-shaped fruiting body of rust fungi in which aeciospores are borne.

Alkaloid Any group of nitrogenous organic bases found in plants with toxic or medicinal properties.

Alternate host One of two hosts required by a pathogen to complete its life-cycle.

Alternative host One of several plant species that are hosts for a given pathogen.

Anamorph The asexual reproductive stage in the life-cycle of a fungus. See imperfect state.

Anthesis The process of flowering which includes the release of pollen.

Appressorium (pl. **appressoria**) A swollen fungal hyphal tip usually associated with adherence to and penetration of the plant surface.

Ascocarp Fruiting body in or on which asci are produced.

Ascospore A sexually- produced spore borne in an ascus.

Ascus (pl. **asci**) A spore sac containing ascospores.

Asexual Vegetative.

Autoecious A rust fungus capable of completing the life-cycle on one host. See heteroecious.

Avirulent Lacking virulence.

Basidiocarp Fruiting body in or on which basidia are produced.

Basidiospore A sexually produced spore borne on a basidium.

Basidium (pl. **basidia**) A club-shaped structure on which sexually produced basidiospores are borne. Sometimes called promycelium.

Biological control The reduction of the amount of inoculum or disease-producing activity of a pathogen accomplished by or through one or more organisms (antagonists) other than man.

Biotroph Organism which is entirely dependent upon another living organism as a source of nutrients (obligate parasite).

Blight A disease characterised by the rapid death of plant tissue.

Broad-spectrum fungicide A fungicide with activity against a wide range of pathogens.

Canker A necrotic often sunken lesion.

Chemotherapy The treatment of disease by chemical means.

Chlamydospore An asexually produced, thick-walled resting spore.

Chlorosis Yellowing of usually green plant tissue.

Circulative virus A virus which passes through the gut wall of the vector into the haemolymph and eventually contaminates the mouthparts via the saliva. See persistent virus.

Cirrus (pl. **cirri**) A gelatinous tendril-like mass of extruded spores.

Cleistothecium (pl. **cleistothecia**) A closed, often spherical ascocarp.

Colonisation The spread of the pathogen in the host tissue away from the initial site of infection and the dependence on the host for nutrients.

Conidiophore A specialised hyphal branch bearing conidia.

Conidium (pl. **conidia**) An asexually produced fungal spore.

Damping-off The rot of seedlings near soil level after emergence (post-emergence) or before emergence (pre-emergence).

Differential variety A variety which gives reactions which distinguish between race-specific isolates of a pathogen.

Dikaryotic Containing two sexually compatible nuclei per cell.

Diploid Having two sets of chromosomes.

Disease A harmful deviation from normal functioning of physiological processes.

Disinfestation The destruction of a pathogen on the surface of the host or in the environment surrounding the host.

Durable resistance Resistance which remains effective in varieties that are extensively cultivated in environments favourable to disease.

Epidemic A progressive increase in the incidence of a particular disease within a defined host population.

Epidemiology The study of factors influencing the development of a disease epidemic.

Epidermis Outer layer of tissue.

Eradicant fungicide A fungicide used to kill existing pathogen infestation. Often referred to as a curative fungicide.

Ergot The sclerotia of the fungal genus *Claviceps*.

Exudate Substance passed from within a plant to the outer surface or into the surrounding medium.

Facultative parasite An organism able to live as a saprophyte or a parasite.

Field resistance Resistance detectable under natural infection in field conditions. See adult plant resistance.

Flagellum (pl. **flagella**) A whip-like organ of motility found on bacteria and zoospores.

Foot rot A disease characterised by a rot of basal stem tissues.

Forma specialis (pl. formae speciales) Strains of a fungal pathogen that are morphologically indistinguishable but pathogenically specialised to different host species

Fumigation Disinfestation by toxic fumes.

Fungicidal Able to kill fungal spores or mycelium.

Fungicide A substance that kills fungal spores or mycelium.

Fungicide resistance A decrease in sensitivity to a fungicide due to selection or mutation following exposure to the compound.

Fungistatic Able to stop fungal growth without killing the fungus.

Gall An abnormal growth or swelling produced as a result of pathogen invasion.

Germ tube The initial hyphal growth from a germinating fungal spore.

Glume The outer bracts of a spikelet in the flowers of cereals and grasses.

Green bridge Living plant material used by pathogens to survive between susceptible hosts.

Haploid Having one complete set of chromosomes.

Haustorium (pl. **haustoria**) A specially developed fungal hyphal branch within a living cell of the host for absorption of food.

Heteroecious A rust fungus that requires two host species to complete its life-cycle.

Heterothallic The condition in which sexual reproduction can only occur between different sexually compatible mycelia (thalli).

Homothallic The condition in which sexual reproduction can occur within the fungal mycelium (thallus).

Honeydew Sugary liquid on plant surfaces often secreted by aphids.

Host An organism harbouring a parasite.

Hybridisation The crossing of two individuals differing in one or more heritable characteristics resulting in the production of a hybrid.

Hyperplasia Abnormal growth associated with increased cell division.

Hypersensitivity A rapid local reaction of plant tissue to attack by a pathogen resulting in the death of tissue around infection sites preventing further spread of infection.

Hypertrophy Abnormal growth associated with cell enlargement.

Hypha (pl. **hyphae**) A tubular thread-like filament of fungal mycelium.

Hyphopodium (pl. **hyphopodia**) A short mycelial branch.

Immune Exempt from infection.

Imperfect state The asexual reproductive stage in the life-cycle of a fungus.

Incubation period The period of time between infection and the appearance of symptoms.

Infection The early stages of pathogen development within a host.

Infection peg A slender hyphal structure penetrating a host cell

Inoculum (pl. **inocula**) Spores or other pathogen parts which can initiate disease.

Intercellular Between cells.

Intracellular Within or through cells.

Karyogamy Fusion of nuclei of two gametes.

Latent period The time between infection and sporulation of the pathogen on the host, or time from the start of a virus vector's feeding period until the vector is able to transmit the virus to healthy plants.

Lesion A localised area of diseased or disordered tissue.

Lodging Breakage of plant stems, especially cereals resulting in tillers falling down.

Major gene resistance Genetic resistance to disease based on one or a few genes.

Monocyclic Having only one cycle of infection during a growing season.

Mosaic Patchy variation of normal green colour. Symptomatic of many virus diseases.

Mould Generic name for fungal growth over a substrate.

Mottle An arrangement of indistinct light and dark areas. Symptomatic of many virus diseases.

Multiline A variety composed of almost genetically identical breeding lines (isogenic)with common agronomic characters, but different major genes for resistance.

Mycelium A mass of hyphae that form the vegetative body of a fungus.

Mycotoxin Toxins produced by fungi which may contaminate foodstuffs.

Necrosis A browning or blackening of cells as they die.

Necrotroph An organism that causes the death of host tissues as it grows through them such that it is always colonising dead substrate.

Node The part of a stem from which a leaf arises.

Non-persistent virus A virus that persists in its vector for a few (usually less than 4) hours at approximately 20°C.

Obligate parasite An organism capable of living only as a parasite (biotroph).

Oospore A sexually produced resting spore of fungi in the class oomycetes.

Pathogen An organism which causes disease.

Pathogenicity The ability to cause disease.

Pathovar (pathotype) Strains of a pathogen, usually a bacterium, which are indistinguishable in physiologic tests but pathogenically specialised to different host species.

Parasite An organism or virus which lives on another living organism (host), obtaining its nutrient supply from the host but conferring no benefit in return.

Perfect State The sexual reproductive stage in the life-cycle of a fungus.

Perithecium (pl. perithecia) A closed flask-shaped ascocarp having an apical hole.

Persistence Time for which a virus vector remains infective after leaving the virus source.

Persistent virus A virus which persists in vector for more than 100 hours and in some cases for the life of the vector.

Phytotoxic Toxic to plants.

Plasmodium (pl. plasmodia) A naked amoeboid multinucleate mass of protoplasm.

Polycyclic Having more than one cycle of infection during a growing season.

Polygenic resistance Genetic resistance to disease based on many genes.

Propagative virus A virus which multiplies in its vector.

Propagule That part of an organism by which it may be dispersed or reproduced.

Prophylaxis Preventative treatment against disease.

Protectant fungicide A fungicide which protects against invasion by a pathogen.

Pseudothecium (pl. pseudothecia) A fruiting body containing asci similar in appearance to a perithecium, but produced in an aggregation of vegetative hyphae.

Pustule A blister-like spore mass breaking through a plant epidermis.

Pycnidium (pl. pycnidia) A flask-shaped or spherical fungal receptacle bearing asexual spores, pycnospores.

Pycnospore An asexual spore produced in a pycnidium.

Race Strains of a fungal or bacterial pathogen that are morphologically or physiologically indistinguishable, but pathogenically specialised to different varieties of a host species.

Race non-specific resistance Resistance to all races of a pathogen.

Race-specific resistance Resistance to some races of a pathogen, but not to others.

Resistant Possessing qualities which prevent or retard the development of a given pathogen.

Resting spore A thick-walled spore that remains dormant for a period of time before germination.

Rhizosphere The zone in soil affected by roots.

Roguing Removal of diseased or unwanted plants from a crop.

Rot Disintegration of tissue.

Saprophyte An organism that lives on dead and decaying material.

Scab A roughened incrustation. A disease in which such lesions form.

Sclerotium (pl. sclerotia) A long-lived compacted mass of vegetatively produced hyphae.

Seedling resistance Resistance detectable at the seedling stage.

Semi-persistent virus A virus which persists in its vector for between 10 and 100 hours.

Senescence Ageing which eventually leads to death.

Sign A pathogen or parts of a pathogen observed on a diseased plant.

Specific (test) weight Weight of a given volume of grain usually expressed in kilograms/hectolitre

Spermatium (pl. **spermatia**) A gamete produced in a spermogonium.

Spermogonium (pl. **spermogonia**) A fruiting body in which gametes (spermatia) are produced.

Spikelet The grouping of the flowers in the inflorescence of cereals and grasses.

Sporangiophore A specialised hyphal branch bearing sporangia.

Sporangiospore A non-motile asexual spore produced in a sporangium.

Sporangium (pl. **sporangia**) A container in which asexual spores are produced. Sometimes functions as a single spore.

Spore A specialised propagative or reproductive body in fungi.

Spreader A substance added to a spray to assist in its even distribution over the target.

Sterilisation The elimination of micro-organisms.

Sticker A substance added to a spray to assist in its adhesion to the target.

Straggling Breakage of a few plant stems, especially cereals resulting in a few tillers falling down.

Stroma (pl. **stromata**) A mass of vegetative hyphae in or on which spores are produced.

Stylet-borne virus A virus which is borne on the stylet of its vector.

Suppressive soil Soil in which a pathogen may persist, but either causes little or no damage or causes disease for a short time and then declines.

Surfactant A surface active material, especially a wetter or spreader used with a spray.

Susceptible Subject to infection.

Symptom A visible change in a host plant as a result of pathogen infection.

Systemic fungicide A fungicide which is absorbed and translocated in the plant.

Systemic infection An infection that spreads throughout the plant from a single infection point.

Take-all decline The decline in the cereal disease take-all after three or four successive cereal crops.

Target spot A lesion consisting of a dark brown circular area containing brown concentric rings. Typical of infection by *Alternaria* spp.

Teleomorph The sexual reproductive stage in a fungal life cycle.

Teliospore A resting spore of rust and smut fungi in which karyogamy occurs.

Tolerant Able to endure infection by a pathogen without showing severe symptoms of disease, or, able to compensate for the effects of disease.

Uredinium (pl. **uredinia**) A fruiting body of the rust fungi in which urediniospores are produced.

Urediniospore An asexual spore of the rust fungi.

Vector An organism which transmits a pathogen, usually a virus.

Virulence The relative ability to cause disease.

Viruliferous A vector which carries and can transmit a virus.

Volunteer plant A self-sown plant, especially cereal.

Whitehead A bleached cereal ear containing little or no grain. Usually a result of attack by stem base or root pathogens, particularly *Gaeumannomyces graminis* (take-all).

Wilt Loss of turgor in plant parts resulting in drooping.

Yellows A plant disease characterised by a general yellowing of tissue and stunting of plants

Zoosporangium (pl. **zoosporangia**) A sporangium containing or producing zoospores.

Zoospore A fungal spore capable of movement in water.

Appendices

Appendix 1

Diagnostic features of the some of the most important cereal diseases

Diagnostic features	Cereal affected	Disease and pathogen
Ear and grain diseases		
Grain replaced by mass of black spores	Wheat Barley Oats Rye	Loose smut (*Ustilago nuda*) Loose smut (*Ustilago nuda*) Loose smut (*Ustilago avenae*) Covered smut (*Ustilago hordei*)
Yellow or brown pustules inside glumes or on awns	Wheat Barley Rye	Rusts (see below)
White mould mainly on surface of glumes	Wheat	Mildew (see below)
Purple–brown tip to glume of still green ear	Wheat	*Septoria nodorum*
Black mould on surface of ripening ear	Wheat Barley Oats Rye	Sooty moulds (*Alternaria* spp. and *Cladosporium* spp.)
Pink and orange spore patches especially at base of spikelet; some bleached spikelets; small dark dots (scab)	Wheat Barley Oats Rye	Ear blight (*Fusarium culmorum*) Scab (*Fusarium graminerum*)
'Whiteheads' – prematurely bleached ripe ears	Wheat Barley Rye	Take all, Eyespot (see below)
Leaf and stem diseases		
White powdery pustules on leaf surface	Wheat Barley Oats Rye	Mildew (*Erysiphe graminis* f.sp. *tritici*) Mildew (*Erysiphe graminis* f.sp. *hordei*) Mildew (*Erysiphe graminis* f.sp. *avenae*) Mildew (*Erysiphe graminis* f.sp. *secalis*)
Yellow pustules usually in stripes	Wheat Barley Rye	Yellow rust (*Puccinia striiformis*, f.sp. *tritici*) Yellow rust (*Puccinia striiformis*, f.sp. *hordei*) Yellow rust (*Puccinia striiformis*, f.sp. *secalis*)
Brown pustules with pale halos scattered at random	Wheat Barley Rye	Brown rust (*Puccinia recondita*, f.sp. *tritici*) Brown rust (*Puccinia recondita*, f.sp. *hordei*) Brown rust (*Puccinia recondita*, f.sp. *secalis*)
Orange pustules often grouped into irregular patches	Oats	Crown rust (*Puccinia coronata*)
Diamond-shaped orange–red pustules followed by black pustules with ruptured epidermis	Wheat Barley Oats Rye	Black stem rust (*Puccinia graminis*, f.sp. *tritici*) Black stem rust (*Puccinia graminis*, f.sp. *hordei*) Black stem rust (*Puccinia graminis*, f.sp. *avenae*) Black stem rust (*Puccinia graminis*, f.sp. *secalis*)
Irregular brown areas with yellow margins occasionally with small brown pycnidia	Wheat Rye	*Septoria nodorum*
Short brown stripes on upper leaves often containing large black pycnidia	Wheat Rye	*Septoria tritici*
Grey 'water soaked' lesions with dark	Barley Rye	Leaf blotch (*Rhynchosporium secalis*)

Diagnostic features	Cereal affected	Disease and pathogen
Leaf and stem diseases (*cont.*)		
Narrow brown stripes on all leaves of plant	Barley	Leaf stripe (*Pyrenophora graminea*)
Yellow, red, or purple discoloration of from leaves tip towards base; discoloration may later turn red or purple	Barley Oats Wheat Rye	Barley yellow dwarf virus (BYDV)
Elongated chlorotic streaks in leaf; stunted and spiky plants	Barley	Barley yellow mosaic virus (BaYMV)
Stem base and root diseases		
Roots blackened, plants stunted in patches	Wheat Barley Oats Rye	Take-all (*Gaumannomyces graminis*; var *graminis*) Take-all (*Gaumannomyces graminis*; var *graminis*) Take-all (*Gaumannomyces graminis*; var *avenae*) Take-all (*Gaumannomyces graminis*; var *graminis*)
Dark brown, eye-shaped lesions below 1st node	Wheat Barley Oats Rye	Eyespot (*Pseudocercosporella herpotrichoides*)
Pale cream, sharply defined irregular lesions above 1st node	Wheat Barley Oats Rye	Sharp eyespot (*Rhizoctonia cerealis*)

Appendix 2

Worldwide distribution and cereal affected by each disease and pathogen

| Common name | Scientific name of pathogen | Occurrence on | | | | Distribution |
		W	B	O	R	
Ear and grain diseases						
Bacteria						
Black chaff (see Bacterial streak)	*Xanthomonas campestris* p.v. *translucens* (syn. *X. translucens*)	**	**	–	*	Worldwide
Fungi						
Black point, Kernel smudge, Sooty mould	*Alternaria, Bipolaris* and *Cladosporium* spp.	*	*	*	*	Worldwide
Bunt, Stinking smut	*Tilletia tritici* (syn. *T. caries*), *T. laevis* (syn. *T. foetida*)	*	–	–	*	Worldwide
Dwarf bunt	*Tilletia controversa*	*	*	–	*	N. America, W Europe, Central Asia, S. America
Karnal bunt	*Tilletia indica* (syn. *Neovossia indica*)	*	–	–	–	India, N. America
Loose smut	*Ustilago tritici*	*	–	–	*	Worldwide
	Ustilago nuda	–	*	–	–	Worldwide
Semi-loose smut, Black loose smut, False loose smut	*Ustilago avenae* (syn. *U. nigra*)	–	*	*	–	Worldwide
Covered smut	*Ustilago hordei*	–	*	*	*	Worldwide
Ergot	*Claviceps purpurea*	*	*	*	*	Worldwide
Head blight, Scab (see Fusarium seedling blight and foot rot)	*Fusarium avenaceum*	*	*	*	*	W Europe
	Fusarium culmorum	*	*	*	*	W Europe
	Fusarium graminearum	*	*	*	*	N. America, S. America, Middle East, Australasia, Asia
	Microdochium nivale	*	*	*	*	W Europe
Leaf and stem diseases						
Bacteria						
Bacterial streak (also known as Black chaff)	*Xanthomonas campestris* p.v. *translucens* (syn. *X. translucens*)	**	**	–	*	Worldwide
Fungi						
Brown rust, Leaf rust	*Puccinia recondita*	*	*	–	*	Worldwide
	Puccinia hordei	–	*	–	–	Worldwide
Cephalosporium stripe	*Cephalosporium gramineum* (syn. *Hymenula cerealis*)	**	*	*	*	N. America, W. Europe, NE. Asia
Crown rust	*Puccinia coronata* f.sp. *avenae*	–	(*)	**	–	Worldwide
Flag smut	*Urocystis agropyri* (syn. *U. tritici*)	**	*	–	–	Worldwide
Halo spot	*Selenophoma donacis; Pseudoseptoria donacis*)	*	*	*	*	Worldwide
Leaf stripe	*Pyrenophora graminea*	–	*	–	–	Worldwide
Net blotch	*Pyrenophora teres* f.sp. *teres* – net form	–	**	*	–	N. America, W. Europe, Middle East
	Pyrenophora teres f.sp. *maculata* – spot form	–	*	*	–	N. America, W. Europe, Middle East
Powdery mildew	*Erysiphe (Blumeria) graminis* f.sp. *tritici*	*	–	–	–	Worldwide
	Erysiphe (Blumeria) graminis f.sp. *hordei*	–	*	–	–	Worldwide
	Erysiphe (Blumeria) graminis f.sp. *avenae*	–	–	*	–	Worldwide
	Erysiphe (Blumeria) graminis f.sp. *secalis*	–	–	–	*	Worldwide

W, Wheat; B, Barley; O, Oats; R, Rye; **, primary economic host; *, economic host; (*), rare.

Leaf and stem diseases (*cont.*)

Fungi (cont.)

Common name of disease	Scientific name of pathogen	Occurrence on				Distribution
		W	B	O	R	
Scald, Leaf blotch	*Rhynchosporium secalis*	–	**	–	*	N. America, W. Europe, Asia, Australasia
Septoria leaf and glume blotch	*Stagonospora nodorum; Phaeosphaeria nodorum*	**	*	*	–	Worldwide
Septoria leaf blotch	*Septoria tritici; Mycosphaerella graminicola*	**	–	–	–	Worldwide
Pink snow mould	*Microdochium nivale* (syn. *Fusarium nivale*); *Monographella nivalis*	*	*	*	*	N. Hemisphere: N. America, Europe, Asia
Snow scald	*Myriosclerotinia borealis* (syn. *Sclerotinia borealis, S. graminearum*)	**	*	–	–	N. Hemisphere: N. America, Europe, Asia
Speckled snow mould	*Typhula ishikariensis*	**	–	–	–	N. Hemisphere: N. America, Europe, Asia
	Typhula idahoensis	**	–	–	*	N. Hemisphere: N. America, Europe, Asia
	Typhula incarnata (syn. *T. itoana*)	**	*	*	*	N. Hemisphere: N. America, Europe, Asia
Spot blotch (see Common root rot)	*Cochliobolus sativus; Bipolaris sorokiniana* (syn. *Helminthosporium sativum, H. sorokinianum*)	**	**	*	*	Worldwide
Stem rust (black)	*Puccinia graminis* f.sp. *tritici*	*	*	–	–	Worldwide
	Puccinia graminis f.sp. *avenae*	–	–	*	–	Worldwide
	Puccinia graminis f.sp. *secalis*	–	–	–	*	Worldwide
Tan spot (yellow leaf spot)	*Pyrenophora tritici-repentis* (syn. *P. trichostoma*); *Drechslera tritici-repentis* (syn. *Helminthosporium tritici-repentis*)	**	*	–	*	Worldwide
Yellow rust, Stripe rust	*Puccinia striiformis* f.sp. *tritici*	**	*	–	–	Worldwide
	Puccinia striiformis f.sp. *hordei*	–	*	–	–	Worldwide
	Puccinia striiformis f.sp. *secalis*	–	–	–	*	Worldwide

Viruses

Common name of disease	Scientific name of pathogen	W	B	O	R	Distribution
Barley stripe mosaic	Barley stripe mosaic virus	*	**	–	*	Worldwide
Barley yellow dwarf	Barley yellow dwarf virus	*	**	*	*	Worldwide
Barley yellow mosaic	Barley yellow mosaic virus; Barley mild mosaic virus	–	*	–	–	W. Europe, Asia
Wheat soil-borne mosaic	Wheat soil-borne mosaic virus	**	*	–	*	N. America, S. America, Middle East, Asia, W. Europe
Wheat streak mosaic	Wheat streak mosaic virus	**	*	*	*	N. America, W. Europe, Asia, Middle East
Wheat yellow mosaic (Wheat spindle streak mosaic)	Wheat yellow mosaic virus; Wheat spindle streak mosaic virus	**	*	*	*	N. America, W. Europe, Asia

Stem base and root diseases

Fungi

Common name of disease	Scientific name of pathogen	W	B	O	R	Distribution
Common root rot, Fusarium seedling blight, Foot rot (also known as Spot blotch)	*Cochliobolus sativus; Bipolaris sorokiniana* (syn. *Helminthosporium sativum, H. sorokinianum*)	** (*)	**	*		Worldwide
Fusarium seedling blight, Foot rot, Dryland foot rot	*Fusarium* spp.	* (*)	*	*		Worldwide
Take-all	*Gaeumannomyces graminis* var. *avenae*					Worldwide
	Gaeumannomyces graminis var. *graminis*	–	–	*		Worldwide
	Gaeumannomyces graminis var. *tritici*	–				Worldwide
Eyespot	*Pseudocercosporella herpotrichoides; Tapesia yallundae*	* (*)	*	–		Worldwide
Sharp eyespot	*Rhizoctonia cerealis; Ceratobasidium cereale*	** (*)	*	–		Europe, N. America, Australasia

W, Wheat; B, Barley; O, Oats; R, Rye; **, primary economic host; *, economic host; (*), rare.

Bibliography

GENERAL BIBLIOGRAPHY

Agrios, G.N. (1988), *Plant Pathology*, 3rd edn, New York, Academic Press.

Bockus, W.W. (1987), Diseases of roots, crowns, and lower stems, in *Wheat and Wheat Improvement*, 2nd edn, Heyne, E.G. (ed.), Agronomy Monograph No. 13, pp 510–527, ASA-CSSA-SSSA, Madison.

Brakke, M.K. (1987) Virus diseases of wheat, in *Wheat and Wheat Improvement*, 2nd edn, Heyne, E.G. (ed.), Agronomy Monograph No. 13, pp 585–624, ASA-CSSA-SSSA, Madison.

Bruehl, G.W. (1967), *Diseases other than rust, smut, and virus, in Wheat and Wheat Improvement*, 2nd edn, Heyne, E.G. (ed.), Agronomy Monograph No. 13, pp 375–510, ASA-CSSA-SSSA, Madison.

Cook, R.J., Polley, R.W., and Thomas, M.R. (1991), Disease-induced losses in winter wheat in England and Wales 1985–1989, *Crop Protect.*, **10**, 504–508.

Cunfer, B.M. (1987), Bacterial and fungal blights of the foliage and heads of wheat, in *Wheat and Wheat Improvement*, 2nd edn, Heyne, E.G. (ed.), Agronomy Monograph No. 13, pp 528–541, ASA-CSSA-SSSA, Madison.

Dickson, J.G. (1956) *Diseases of Field Crops*, 2nd edn, New York, McGraw-Hill Book Co.

Fischer, G.W. and Holton, C.S. (1957), *Biology and Control of the Smut Fungi*, New York, Ronald Press Co.

Gair, R., Jenkins, J.E.E., and Lester, E. (1987), *Cereal Pests and Diseases*, Ipswich, Farming Press.

Gareth Jones, D. and Clifford, B.C. (1983), *Cereal Diseases: Their Pathology and Control*, Chichester, Wiley-Interscience.

Holton, C.S., Hoffmann, J.A., and Duran, R. (1968), Variation in the smut fungi, *Ann. Rev. Phytopathol.*, **6**, 213–242.

Mathre, D.E. (ed.) (1982), *Compendium of Barley Diseases*, St Paul, American Phytopathological Society.

McKinney, H.H. (1967), *Virus diseases, in Wheat and Wheat Improvement*, 2nd edn, Heyne, E.G. (ed.), Agronomy Monograph No. 13, pp 355–374, ASA-CSSA-SSSA, Madison.

Parry, D.W. (1990), *Plant Pathology in Agriculture*, Cambridge, Cambridge University Press.

Polley, R.W. and Thomas, M.R. (1991), Survey of diseases of winter wheat in England and Wales, 1976–1988, *Ann. Appl. Biol.*, **119**, 1–20.

Polley, R.W., King, J.E., and Jenkins, J.E.E. (1993), Surveys of diseases of spring barley in England and Wales, 1976–1980, *Ann. Appl. Biol.*, **123**, 271–285.

Polley, R.W., Thomas, M.R., Slough, J.E., and Bradshaw, N.J. (1993), Surveys of diseases of winter barley in England and Wales, 1981–1991, *Ann. Appl. Biol.*, **123**, 287–307.

Schafer, J.F. (1987), Rusts, smuts, and powdery mildew, in *Wheat and Wheat Improvement*, 2nd edn, Heyne, E.G. (ed.), Agronomy Monograph No. 13, pp 524–584, ASA-CSSA-SSSA, Madison.

Smith, I.M., Dunez, J., Phillips, H.H., Lelliott, R.A., and Archer, S.A. (eds) (1988), *European Handbook of Plant Diseases*, Oxford, Blackwell Scientific Publications.

Sprague, R. (1950), *Diseases of Cereals and Grasses in North America (Fungi, Except Smuts and Rusts)*, New York, Ronald Press Co.

Wiese, M.V. (1987), *Compendium of Wheat Diseases*, 2nd edn, St Paul, American Phytopathological Society.

EAR AND GRAIN DISEASES

Boosalis, M.G. (1952). The epidemiology of *Xanthomonas translucens* (J.J. and R.) Dowson on cereals and grasses, *Phytopathol.*, **42**, 387–395.

Culshaw, F., Cook, R.J., Magan, N., and Evans, E.J. (1988), *Blackpoint of Wheat*, Home Grown Cereals Authority Research Review No. 7, London, HGCA.

Cunfer, B.M. and Scolari, B.L. (1982), *Xanthomonas campestris* pv. *translucens* on triticale and other small grains, *Phytopathol.*, **72**, 683–686.

Durán, R. (1987), *Ustilaginales of Mexico. Taxonomy, Symptomatology, Spore Germination, and Basidial Cytology*, Pullman, Dept Plant Pathology, Washington State University.

Forster, R.L. and Schaad, N.W. (1988), Control of black chaff of wheat with seed treatment and a foundation seed health program, *Plant Dis.*, **72**, 935–938.

Fourest, E., Rehms, L.D., Sands, D.C., Bjarko, M., and Lund, R.E. (1990), Eradication of *Xanthomonas campestris* pv. *translucens* from barley seed with dry heat treatments, *Plant Dis.*, **74**, 816–818.

Fuentes-Davila, G. and Rajaram, S. (1994),

Sources of resistance to *Tilletia indica* in wheat, *Crop Protect.*, **13**, 20–24.

Gill, B.S., Randhawa, A.S., Aujla, S.S., Dhaliwal, H.S., Grewal, A.S., and Sharma, I. (1981), Breeding wheat varieties resistant to karnal bunt, *Crop Improv.*, **82**, 73–80.

Grey, W.E., Mathre, D.E., Hoffmann, J.A., Powelson, R.L., and Fernández, J.A. (1986), Importance of seedborne *Tilletia controversa* for infection of winter wheat and its relationship to international commerce, *Plant Dis.*, **70**, 122–125.

Hoffmann, J.A. (1982), Bunt of wheat, *Plant Dis.*, **66**, 979–986.

Hoffmann, J.A. and Waldher, J.T. (1981), Chemical seed treatments for controlling seedborne and soilborne common bunt of wheat, *Plant Dis.*, **65**, 256–259.

Jones, J.P. and Collins, F.C. (1971) Control of loose smut of wheat with carboxin and benomyl, *Plant Dis. Reporter*, **55**, 1053–1055.

Joshi, L.M., Singh, D.V., Srivastava, K.D., and Wilcoxson, R.D. (1983) Karnal bunt: A minor disease that is now a threat to wheat, *Bot. Rev.*, **49**, 309–330.

Kavanagh, T. (1961), Temperature in relation to loose smut in barley and wheat, *Phytopathol.*, **51**, 189–193.

Line, R.F. (1993), Common bunt, in *Seed-Borne Diseases and Seed Health Testing of Wheat*, Mathur, S.B. and Cunfer, B.M. (eds), pp 45–52, Hellerup, Danish Government.

Line, R.F. (1993), Dwarf bunt, in *Seed-Borne Diseases and Seed Health Testing of Wheat*, Mathur, S.B. and Cunfer, B.M. (eds), pp 23–29, Hellerup, Danish Government.

Milus, E.A. and Mirlohi, A.F. (1994), Use of disease reactions to identify resistance in wheat to bacterial streak, *Plant Dis.*, **78**, 157–161.

Moffett, M.J. and Croft, B.J. (1983), Xanthomonas, in *Plant Bacterial Diseases: A Diagnostic Guide*, Fahy, P.C. and Persley, G.J., (eds), pp 189–228, Sydney, Academic Press.

Nielsen, J. (1977), A collection of cultivars of oats immune or highly resistant to smut, *Can. J. Plant Sci.*, **57**, 199–212.

Nielsen, J. (1983), Spring wheats immune or highly resistant to *Ustilago tritici*, *Plant Dis.*, **67**, 860–863.

Nielsen, J. (1987), Races of *Ustilago tritici* and techniques for their study, *Can. J. Plant Pathol.*, **9**, 91–105.

Nielsen, J. (1993), Host specificity of *Ustilago avenae* and *U. hordei* on eight species of *Avena*, *Can. J. Plant Pathol.*, **15**, 14–16.

Nielsen, J. and Tikhomirov, V. (1993), Races of *Ustilago tritici* identified in field collections from Eastern Siberia using Canadian and Soviet differentials, *Can. J. Plant Pathol.*, **15**, 193–200.

Parry, D.W., Jenkinson, P., and McLeod, L. (1995), *Fusarium* ear blight (scab) in small grain cereals – a review, *Plant Pathol.*, **44**, 207–238.

Purdy, L.H., Kendrick, E.L., Hoffmann, J.A., and Holton, C.S. (1963), Dwarf bunt of wheat, *Ann. Rev. Micro.*, **17**, 199–222.

Schaad, N.W. and Forster, R. (1993), *Black chaff*, in *Seed-Borne Diseases and Seed Health Testing of Wheat*, Mathur, S.B. and Cunfer, B.M. (eds), pp 129–136, Hellerup, Danish Government.

Singh, B.B., Aujula, S.S., and Sharma, I. (1993), Integrated management of wheat Karnal bunt, *Int. J. Pest Manage.*, **39**, 431–434.

Snijders, C.H.A. (1990), Genetic variation for resistance to *Fusarium* headblight in bread wheat, *Euphytica*, **50**, 171–179.

Thomas, P.L. (1981), Distinguishing between the loose smuts of barley, *Plant Dis.*, **65**, 834.

Wilcoxson, R.D. and Stuthman, D.D. (1993), Evaluation of oats for resistance to loose smut, *Plant Dis.*, **77**, 818–821.

LEAF AND STEM DISEASES

Adams, M.J., Swaby, A.G., and Jones, P. (1988), Confirmation of the transmission of barley yellow mosaic virus (BaYMV) by the fungus *Polymyxa graminis*, *Ann. Appl. Biol.*, **112**, 133-141.

Adee, E.A. and Pfender, W.F. (1989), The effect of primary inoculum level of *Pyrenophora tritici-repentis* on tan spot epidemic development in wheat. *Phytopathol.*, **79**, 873–877.

Allan, R.E. (1976), Flag smut reaction in wheat – its genetic control and association with other traits, *Crop Sci.*, **16**, 685–687.

Anderegg, J.C. and Murray, T.D. (1988), Influence of soil matric potential and soil pH on Cephalosporium stripe of winter wheat in the greenhouse, *Plant Dis.*, **72**, 1011–1016.

Armitage, C.R., Hunger, R.M., Sherwood, J.L., and Weeks, D.L. (1990), Relationship between development of hard red winter wheat and expression of resistance to wheat soilborne mosaic virus, *Plant Dis.*, **74**, 356–359.

Bissonette, S.M., D'Arcy, C.J., Pedersen, W.L (1994), Yield loss in two spring oat cultivars due to *Puccinia coronata* f. sp. *avenae* in the presence or absence of barley yellow dwarf virus, *Phytopathol.*, **84**, 363–371.

Bockus, W.W. (1993), Evaluation of foliar fungicides on winter wheat for control of tan spot and leaf rust, 1992, *Fungicide Nematicide Tests*, **48**, 225.

Bockus, W.W. and Claassen, M.M. (1992), Effects

of crop rotation and residue management practices on severity of tan spot of winter wheat, *Plant Dis.*, **76**, 633–636.

Bockus, W.W., O'Connor, J.P., and Raymond, P.J. (1983), Effect of residue management method on incidence of Cephalosporium stripe under continuous winter wheat production, *Plant Dis.*, **67**, 1323–1324.

Boosalis, M.G. (1952), The epidemiology of *Xanthomonas translucens* (J.J. and R.) Dowson on cereals and grasses, *Phytopathol.*, **42**, 387–395.

Bruehl, G.W. (1957), Cephalosporium stripe disease of winter wheat in Washington, *Plant Dis. Reporter*, **52**, 590–594.

Bruehl, G.W. (1982), Developing wheats resistant to snow mold in Washington state, *Plant Dis.*, **66**, 1090–1095.

Bruehl, G.W., Sprague, R., Fischer, W.R., Nagamitsu, M., Nelson, W.L., and Vogel, O.A. (1966), *Snow Molds of Winter Wheat in Washington*, Pullman, Washington Agricultural Experimental Station Bulletin, No. 677, 21 pp.

Bruehl, G.W., Murray, T.D., and Allan, R.E. (1986), Resistance of winter wheats to Cephalosporium stripe in the field, *Plant Dis.*, **70**, 314–316.

Campbell, L.G., Heyne, E.G., Gronau, D.M., and Niblett, C. (1975), Effect of soilborne wheat mosaic virus on wheat yield, *Plant Dis. Reporter*, **51**, 1005–1008.

Carroll, T.W. (1980), Barley stripe mosaic virus: its economic importance and control in Montana, *Plant Dis.*, **64**, 136–140.

Christian, M.L. and Willis, W.G. (1993), Survival of wheat streak mosaic virus in grass hosts in Kansas from wheat harvest to fall wheat emergence, *Plant Dis.*, **77**, 239–242.

Conner, R.L., Lindwall, C.W., and Atkinson, T.G. (1987), Influence of minimum tillage on severity of common root rot in wheat, *Can. J. Plant Pathol.*, **9**, 56–58.

Conner, R.L., Whelan, E.D.P., and MacDonald, M.D. (1989), Identification of sources of resistance to common root rot in wheat-alien amphiploid and chromosome substitution lines, *Crop Sci.*, **29**, 916–919.

Cook, R.J. (1968), Ecology and possible significance of perithecia of *Calonectria nivalis* in the Pacific Northwest, Phytopathol., 58, 702–703.

Cook, R.J. (1981), Fusarium diseases of wheat and other small grains in North America, in *Fusarium: Diseases, Biology, and Taxonomy*, Nelson, P.E., Ioussoun, T.A., and Cook, R.J. (eds), pp 39–52, University Park, The Pennsylvania State University Press.

Cunfer, B.M., Demski, J.W., and Bays, D.C. (1988), Reduction in plant development, yield, and grain quality associated with wheat

spindle streak mosaic virus, *Phytopathol.*, **78**, 198–204.

Cunfer, B.M. and Bruehl, G.W. (1973), Role of basidiospores as propagules and observations on sporophores of *Typhula idahoensis*, *Phytopathol.*, **63**, 115–120.

Cunfer, B.M. and Scolari, B. L. (1982), *Xanthomonas campestris* pv. *translucens* on triticale and other small grains, *Phytopathol.*, **72**, 683–686.

da Luz, W.C. and Bergstrom, G.C. (1986), Effect of temperature on tan spot development in spring wheat cultivars differing in resistance, *Can. J. Plant Pathol.*, **8**, 451–454.

Dill-Macky, R., Rees, R.G., and Platz, G.J. (1990), Stem rust epidemics and their effects on grain yield and quality in Australian barley cultivars, *Aust. J. Agric. Res.*, **41**, 1057–1063.

Forster, R.L. and Schaad, N.W. (1988), Control of black chaff of wheat with seed treatment and a foundation seed health program, *Plant Dis.*, **72**, 935–938.

Fourest, E., Rehms, L.D., Sands, D.C., Bjarko, M., and Lund, R.E. (1990), Eradication of *Xanthomonas campestris* pv. *translucens* from barley seed with dry heat treatments, *Plant Dis.*, **74**, 816–818.

Goos, R.J., Johnston, B.E., and Stack, R.W. (1989), Effect of potassium chloride, imazalil, and method of imazalil application on barley infected with common root rot, *Can. J. Plant Sci.*, **69**, 437–444.

Griffey, C.A., Das, M.K., Baldwin, R.E., Waldenmaier, C.M. (1994), Yield losses in winter barley resulting from a new race of *Puccinia hordei* in North America, *Plant Dis.*, **78**, 256–260.

Hosford, R.M., Jr (1981), *Tan Spot of Wheat and Related Diseases Workshop*, Fargo, North Dakota State University, 116 pp.

Hosford, R.M., Jr, Jordahl, J.G., and Hammond, J.J. (1990), Effect of wheat genotype, leaf position, growth stage, fungal isolate, and wet period on tan spot lesions, *Plant Dis.*, **74**, 385–390.

Hunger, R.M., Armitage, C.R., and Sherwood, J.L. (1989), Effects of wheat soilborne mosaic virus on hard red winter wheat, *Plant Dis.*, **73**, 949–952.

Irwin, M.E. and Thresh, J.M. (1990), *Epidemiology of Barley Yellow Dwarf: A Study in Ecological Complexity*, Annual Review of Phytopathology, Vol. 28, Cook, R.J., Zentmyer, G.A., and Cowling, E.B. (eds), pp 393–424, Palo Alto, Annual Reviews Inc.

Jamalainen, E.A. (1949), Overwintering of *Gramineae* plants and parasitic fungi. I. *Sclerotinia borealis* Bubák & Vleugel, *J. Sci. Agric. Soc. Finland*, **21**, 125–142.

Jamalainen, E.A. (1974), Resistance in winter cereals and grasses to low-temperature parasitic fungi. *Ann. Rev. Phytopathol.*, **12**, 281–302.

Johnston, H.W. (1976), Influence of spring seeding date on yield loss from root rot of barley, *Can. J. Plant Sci.*, **56**, 741–743.

Johnston, R.H. and Mathre, D.E. (1972), Effect of infection by *Cephalosporium gramineum* on winter wheat, *Crop Sci.*, **12**, 817–819.

Junger, R.M., Sherwood, J.L., Evans, C.K., and Montana, J.R. (1992), Effects of planting date and inoculation date on severity of wheat streak mosaic in hard red winter wheat cultivars, *Plant Dis.*, **76**, 1056–1060.

King, J.E., Cook, R.J., and Melville, S.C. (1983), A review of *Septoria* diseases of wheat and barley, *Ann. Appl. Biol.*, **103**, 345–373.

Kokko, E.G., Conner, R.L., Kozub, G.C, and Lee, B. (1993), Quantification by image analysis of subcrown internode discoloration in wheat caused by common root rot, *Phytopathol.*, **83**, 976–981.

Larsen, H.J., Brakke, M.K., and Langenberg, W.G. (1985), Relationship between wheat streak mosaic virus and soilborne wheat mosaic virus infection, disease resistance, and early growth of winter wheat, *Plant Dis.*, **69**, 857–862.

Lawton, M.B. and Burpee, L.L. (1990), Seed treatments for typhula blight and pink snow mold of winter wheat and relationships among disease intensity, crop recovery, and yield, *Can. J. Plant Pathol.*, **12**, 63–74.

Line, R.F. (1993), Flag smut, in *Seed-Borne Diseases and Seed Health Testing of Wheat*, Mathur, S.B. and Cunfer, B.M. (eds), pp 53–57, Hellerup, Danish Government.

Long, D.L., Roelfs, A.P., Leonard, K.J., and Roberts, J.J. (1994), Virulence and diversity of *Puccinia recondita* f. sp. *tritici* in the United States in 1992, *Plant Dis.*, **78**, 901–906.

Mathre, D.E., Johnston, R.H., and Martin, J.M. (1985), Sources of resistance to *Cephalosporium gramineum* in *Triticum* and *Agropyron* species, *Euphytica*, **34**, 419–424.

McBeath, J.H. (1985), Pink snow mold on winter cereals and lawn grasses in Alaska, *Plant Dis.*, **69**, 722–723.

McBeath, J.H., Drew Smith, J., and Tronsmo, A. M (1993), Pink snow mold, leaf blotch and ear blight, in *Seed-Borne Diseases and Seed Health Testing of Wheat*, Mathur, S.B. and Cunfer, B.M. (eds), pp 95–103, Hellerup, Danish Government.

McKinney, H.H. (1923), Influence of soil temperature and moisture on infection of wheat seedlings by *Helminthosporium sativum*, *J. Agric. Res.*, **26**, 195–218.

Mehta, Y.R. and Igarashi, S. (1985), Chemical control measures for the major diseases of wheat, with special attention to spot blotch, in *Wheats for More Tropical Environments, The Proceedings of an International Symposium*, pp. 196–200, Copenhagen, CIMMYT.

Mehta, Y.R. (1993), Spot blotch, in *Seed-Borne Diseases and Seed Health Testing of Wheat*, Mathur, S.B. and Cunfer, B.M. (eds), pp 105–112, Hellerup, Danish Government.

Miller, N.R., Bergstrom, G.C., Sorrells, M.E., and Cox, W.J. (1990), Effect of wheat spindle streak mosaic on yield of winter wheat in New York, *Phytopathol.*, **82**, 852–857.

Milus, E.A. (1994), Effects of leaf rust and Septoria leaf blotch on yield and test weight of wheat in Arkansas, *Plant Dis.*, **78**, 55–59.

Milus, E.A. and Mirlohi, A.F. (1994), Use of disease reactions to identify resistance in wheat to bacterial streak, *Plant Dis.*, **78**, 157–161.

Moffett, M.J. and Croft, B.J. (1983), *Xanthomonas*, in *Plant Bacterial Diseases. A Diagnostic Guide*, Fahy, P.C. and Persley, G.J. (eds), pp 189–228, Sydney, Academic Press.

Murphy, F.A., Fauquet, CM, Mayo, MA, Jarvis, AW, Ghabrial, SA, Summers, M.D., Martelli, G.P., and Bishop, D.H.L. (eds) (1995), *Sixth Report of the International Committee on Taxonomy of Viruses*, Archives of Virology, New York, Springer Verlag.

Pfender, W.F., Zhang, W., and Nus, A. (1993), Biological control to reduce inoculum of the tan spot pathogen *Pyrenophora tritici-repentis* in surface-borne residues of wheat fields, *Phytopathol.*, **83**, 371–375.

Purdy, L.H. (1965), Flag smut of wheat, *Bot. Rev.*, **31**, 565–606.

Rao, A.S. and Brakke, M.K. (1969), Relation of soil-borne wheat mosaic virus and its fungal vector, *Polymyxa graminis*, *Phytopathol.*, **59**, 581–587.

Raemakers, R. (1985), Chemical control of *Helminthosporium sativum* on rainfed wheat in Zambia, in *Wheats for More Tropical Environments, The Proceedings of an International Symposium*, pp. 201–203, Mexico City, CIMMYT.

Raymond, P.J. and Bockus, W.W. (1984), Effect of seeding date of winter wheat on incidence, severity, and yield loss caused by Cephalosporium stripe in Kansas, *Plant Dis.*, **68**, 665–667.

Roelfs, A.P. (1982), Effects of barberry eradication on stem rust in the United States, *Plant Dis.*, **66**, 177–181.

Roelfs, A.P. (1985), Epidemiology of the cereal rusts in North America, *Can. J. Plant Pathol.*, **11**, 86–90.

Schaad, N.W. and Forster, R. (1993), Black chaff,

in *Seed-Borne Diseases and Seed Health Testing of Wheat*, Mathur, S.B. and Cunfer, B.M. (eds), pp 129–136, Hellerup, Danish Government.

Schilder, A. and Bergstrom, G.C. (1993), Tan spot, in *Seed-Borne Diseases and Seed Health Testing of Wheat*, Mathur, S.B. and Cunfer, B.M. (eds), pp 113–122, Hellerup, Danish Government.

Schneider, E.F. and Seaman, W.L. (1987), Occurrence of *Myriosclerotinia borealis* on winter cereals in Ontario, *Can. Plant Dis. Survey*, **67**, 1–2.

Shabeer, A. and Bockus, W.W. (1988), Tan spot effect on yield and yield components relative to growth stage in winter wheat, *Plant Dis.*, **72**, 599–602.

Slykhuis, J.K. (1970), Factors determining the development of wheat spindle streak mosaic caused by a soil-borne virus in Ontario, *Phytopathol.*, **60**, 319–331.

Slykhuis, J.K. (1976), Virus and virus-like diseases of cereal crops, *Ann. Rev. Phytopathol.*, **14**, 189–210.

Slykhuis, J.T. (1967), Virus diseases of cereals, *Rev. Appl. Mycol.*, **46**, 401–429.

Smith, J.D. (1981), Snow molds of winter cereals: guide for diagnosis, culture, and pathogenicity. *Can. J. Plant Pathol.*, **3**, 15–25.

Sommerfeld, M.L., Gildow, F.E., and Frank, J.A. (1993), Effects of single or double infections with *Helminthosporium avenea* and barley yellow dwarf virus on yield components of oats. *Plant Dis.*, **77**, 741–744.

Sprague, R., Fischer, W.R., and Figaro, P. (1961), Another sclerotial disease of winter wheat in Washington, *Phytopathol.*, **51**, 334–336.

Tomiyama, K. (1955), *Studies on the Snow Blight Diseases of Winter Cereals*, Hokkaido National Agricultural Experimental Station, Rep. No. 47, 234 pp.

Tomiyama, K. (1961), Snow blight of winter cereals in Japan, *Recent Advances in Botany*, Toronto, University of Toronto Press, pp 549–552.

STEM BASE AND ROOT DISEASES

Asher, M.J.C. and Shipton, P.J. (eds) (1981), *Biology and Control of Take-All*, London, Academic Press.

Conner, R.L., Lindwall, C.W., and Atkinson, T.G. (1987), Influence of minimum tillage on severity of common root rot in wheat, *Can. J. Plant Pathol.*, **9**, 56–58.

Conner, R.L., Whelan, E.D.P., and MacDonald, M.D. (1989), Identification of sources of resistance to common root rot in wheat-alien amphiploid and chromosome substitution lines, *Crop Sci.*, **29**, 916–919.

Fitt, B.D.L. (1988), *Eyespot Disease of Cereals*, Home Grown Cereals Authority Research Review No. 1, London, HGCA.

Goos, R.J., Johnston, B.E., and Stack, R.W. (1989), Effect of potassium chloride, imazalil, and method of imazalil application on barley infected with common root rot, *Can. J. Plant Sci.*, **69**, 437–444.

Jenkins, J.E.E., Clark, W.S., and Buckle, A.E. (1988), *Fusarium Diseases of Cereals*, Home Grown Cereals Authority Research Review No. 4, London, HGCA.

Jenkinson, P. and Parry, D.W. (1994), Isolation of *Fusarium* species from common broad leaved weeds and their pathogenicity to winter wheat, *Mycol. Res.*, **98**, 506–510.

Johnston, H.W. (1976), Influence of spring seeding date on yield loss from root rot of barley, *Can. J. Plant Sci.*, **56**, 741–743.

Jones, D.R. (1994), Evaluation of fungicides for control of eyespot disease and yield loss relationships in winter wheat, *Plant Pathol.*, **43**, 831–846.

Kokko, E.G., Conner, R.L., Kozub, G.C., and Lee, B. (1993), Quantification by image analysis of subcrown internode discoloration in wheat caused by common root rot, *Phytopathol.*, **83**, 976–981.

McKinney, H.H. (1923), Influence of soil temperature and moisture on infection of wheat seedlings by *Helminthosporium sativum*, *J. Agric. Res.*, **26**, 195–218.

Mehta, Y.R. (1993), Spot blotch, in *Seed-Borne Diseases and Seed Health Testing of Wheat*, Mathur, S.B. and Cunfer, B.M. (eds), pp 105–112, Hellerup, Danish Government.

Nicholson, P., Rezanoor, H.N., and Hollins, T.W. (1991), Occurrence of *Tapesia yallundae* apothecia on field and laboratory-inoculated material and evidence for recombination between isolates, *Plant Pathol.*, **40**, 626–634.

Index

Bold numbers indicate illustrations.